$5.00
Paperback

PLACER MINER'S MANUAL

Volume 1

This volume contains the complete Part 1 of PLACER MINING IN THE WESTERN UNITED STATES, by E. D. Gardner and C. H. Johnson of the U.S. Bureau of Mines. All of the invaluable, original annotations are included and Nothing has been omitted. Additional editorial and technical material has been provided by Karl von Mueller.

Contents: History of placer mining, geology, sampling, surf washers, classification, panning, sluicing, dry washers, a weight conversion table, tables, illustrations, and maps.

By Karl von Mueller

Library of Congress Catalog No. 71-165766

ISBN 0-89316-611-1 Paperback
LSBN 0-89316-612-X Hardcover

First Printing 1971
Second Printing 1972
Third Printing 1980 Revised

Printed in the United States of America

Published by: Exanimo Press (Gazette Publications)
P.O. Box 18
Segundo, Colorado 81070 U.S.A.

Published from the little one-man printery
in the old ghosttown of Segundo, Colorado
on the historic Purgatoire river
in the foothills of the Sangre de Cristo mountains

Authors, lecturers, and researchers are invited to use parts of this book in order to further disseminate the technical information that is printed herein, provided that due credit is given to authors E. D. Gardner, C. H. Johnson, and Karl von Mueller.

If you would like adance announcements of future books, please send your request to: Exanimo Press
Box 18
Segundo, Colorado 81070.

FOREWORD

Placer Mining in the Western United States, in 3 volumes, were probably the most helpful and informative books on mining that were ever published by the United States government. They were a provident aid and 'bible' for the inexperienced men and women who took to prospecting and mining during the great depression of the 1930s. By the time of the Great War, these fine books had gone out of print and they were never reprinted.

Volume 1, or Part 1, consisted of general information, hand-shoveling, and ground sluicing. It was designated Information Circular 6786 and was published in September 1934 by the U.S. Bureau of Mines.

Volume 2, or Part 2, consisted of hydraulicking, treatment of placer concentrates, and marketing of gold. It was designated Information Circular 6787 and was published in October 1934 by the U.S. Bureau of Mines.

Volume 3, or Part 3, consisted of dredging and other forms of mechanical handling of gravel and drift mining. It was designated Information Circular 6788 and was published in February 1935 by the U.S. Bureau of Mines.

The compilers and authors of these monumental books were E. D. Gardner who was supervising engineer of the USBM Southwest Experiment Station at Tucson, Arizona, and C. H. Johnson who was assistant mining engineer at the same station.

The incredible amount of everyday information that is packed into these three books provides a source of information and a tool that is valuable to the most experienced miner and to the rank beginner.

There is an interesting story behind the original reprinting of these books by Exanimo Press. When I first acquired the National PROSPECTOR'S Gazette in 1954, I requested permission to republish these three books in serial form in the NPG. After a delay of several months, I received permission from the U. S. Bureau of Mines to serialize them in the NPG. A few days later, I received an airmail letter from the Bureau of Mines that did NOT revoke the previous permission to serialize the books but it did express the opinion of somebody in the Bureau that it would not be in the best interests of the Bureau of Mines to serialize these books. I acceded to their wishes and had long since forgotten about the matter.

Along about 1970, a full-time prospector and treasure hunter was visiting at the Exanimo Establishment in Segundo, Colorado (now at Ames, Nebraska 68621). In the course of our visit, he showed me a set of these books that were dog-eared and in poor condition. He had payed a bookfinder $45.00 for the three books. I told him that I owned two sets of these books, one in mint condition, and that I had once considered running the entire contents of the three books in the Prospector's Gazette as a long-running serial. Other prospectors in the store surmised that they would buy a set at almost any price if they could find them. I was asked what it would cost to reprint them and I estimated around $4.00 per copy. They, and others, planted seads of encouragement with me, so I approached the Bureau of Mines again.

I initiated negotiations by asking whether or not there were any plans to reprint these valuable Information Circulars. While waiting for a reply, I set out my poorest copies of these books and asked other prospectors what they thought of them and if they would be of any use to them. Everybody that looked through or read parts of them said they would buy the complete set if I would manage to reprint them. The die was cast.

By the time that a reply was received, I was checking printing and binding costs and they were not at all encouraging. Nevertheless, a few telephone calls set up permission and authority to reprint these books, but a formal request in writing was required. This was submitted and I started to touch-up the broken and missing characters on every page and to ink-in the lines on defective illustrations. By the time that formal permission was received, the books were ready for reproduction, but another problem had arisen.

Estimates and bids by commercial printers pushed the retail cost of each book to $5.00 or more, and this was beyond the reach of many prospectors, small miners, and mining enthusiasts. The answer was to print them myself. I ordered enough paper for 5,000 copies and specified that the paper was to be punched so that it would fit into ordinary 3-ring binders. While waiting for delivery of the paper, I sent the books to an offset platemaking outfit and had negatives and plates made. The day the truckload of paper arrived, I started printing and what a job it was. After printing the pages had to be collated into books and then bound with a removable cover. Our supply of books was exhausted within a year and I simply did not have enough horsepower to print them again.

When the current gold and silver craze began erupting in late 1979, I had just completed writing and publishing the Gold Dredger's Handbook, Second Edition, and the requests for it were coming in at an unexpected rate. The inquries about the Placer Miner's Manuals gave us more mail to answer than we could handle. Paul Tainter of Exanimo Establishment ordered 1,000 copies of the dredging book to start off with and pleaded that I give serious consideration to reprinting the Placer Miner's Manuals.

At that time, I was trying to finish up another book, Vibrating Gold Concentrators, so I decided that when this book was finished, I would consent to reprint the placer books. I was appalled when I started to investigate printing costs but Mr. Tainter, above all others, assured me that he would highlight the book in all of his catalogs and circulars so that the initial printing costs could be amortized as soon as possible. Mr. Tainter's address is Exanimo Establishment, Box 448, Fremont, Nebraska 68025 in case you want to get a catalog of treasure hunting, prospecting, and mining books, equipment, and supplies. He is one of the most honest, considerate, and helpful people I have ever done business with and I commend him to you without reservation of any sort.

BOOKS

Quite often, the most informative part of a book is the bibliography that is, essentially, the anticlimax of the book. It is relatively common knowledge that many books are written on technical subjects by authors who have no right or business engaging in the themes they pursue. Most writers write for glory or the dollar but there are those who write for the lively love of writing or to help people, either by amusing or entertaining them or by providing helpful information.

Here are some books that I can recommend very highly and they are worth every penny of their cost. I am listing the cost of the books that I have at the time that I bought them, so you can figure the prices are much higher now.

Handbook for the Alaskan Prospector, by Ernest Wolff. Mineral Research Laboratory, 1969. xvii-460 pages, 19 chapters, 3 appendices, bibliography, list of illustrations, addendum, index. $5.95. The author covers every facet of mining thoroughly and even though this book relates to Alaskan prospecting and mining, much of its information can be used around the world. Now out-of-print, if you find a copy in a bookstore, buy it!

The Gold Hunter's Field Book, by Jay Ellis Ransom. Harper & Row, 1975. xiv-367 pages, 3 parts are headed 'Introduction to Gold Hunting, Where to Find Gold in the United States, and Where to Find Gold in Canada. A good index is in the back of the book. $9.95. The author knows his business and has prepared a masterpiece for the prospector. In many cases, he has listed the locations down to the range and township and in practically all locations, he gives the names of the streams and their location.

Mineral Facts and Problems (Bureau of Mines Bulletin 630), by Staff, U.S. Bureau of Mines. Government Printing Office or Superintendent of Documents, $6.75. 1118 pages. Covers all commercial minerals with mining technology, treatment and processes, and problems.

Principal Gold-producing Districts in the United States (Geological Survey Professional Paper 610), by A. H. Kaschmann and M. H. Bergdahl. v-283 pages, contents, abstract, introduction, list of illustrations, bibliography, index. Superintendent of Documents, $4.75. Records the history and geology of the mining districts in 21 states.

Dictionary of Mining, Mineral, and Related Terms, by Staff, U. S. Bureau of Mines. Superintendent of Documents, $15.00. vii-1269 pages. This almost a mandatory reference work when you get into some of the professional and technical papers and documentation of mining.

You may find some or all of these books in large bookshops like B. Dalton, Pickwick, Baker & Taylor, Canterbury, Bacon, Countts, Ulbricks, or Walden. If they cannot supply you, I noticed that Exanimo Books, Box 448, Fremont, Nebraska 68025 had all of these titles in stock in April 1980. They had less than a half-dozen copies of the Alaskan Prospector, and one or two each of the government books. I would suggest that you send for their book bulletin which gives their latest large stock of books. Even if you have a small local bookshop, it will not hurt to check it as they might have one or more of the titles listed. There are, of course, hundreds of other books dealing with mining in and out of print and you can often run into some gems that were published in the last century at low prices.

PERIODICALS

In the mining world, everybody has their favorite magazine. Some of these periodicals contain highly technical articles and new reports that are over the head of the ordinary prospector. Some are filled with more advertising than will meet your interest. Some are downright newsy and some of them are oriented for the frugal miner.

Here is a list of some of those that I am sure will interest the average miner and prospector:

Western Prospector & Miner, Box 146, Tombstone, Arizona 85638. Monthly, $5 per year. A fascinating paper that publishes western mining new and reports on the shenanigans of the bureaucrats. Recommended highly.

Pay Dirt, Drawer 48, Bisbee, Arizona 85603. Monthly, $7 per year. Reports on Arizona mining, chiefly. Popular with southwestern miners.

National Prospectors Gazette, Miner's Hall, Ames, Nebraska 68621. Bimonthly, $5 per year. Combined with the Treasure Hunter's News. The maverick of the adventure world. Regular columns include book review, recipes, tips for prospectors, dowsing, Concentrates on money-, time-, and labor-saving ideas.

The Mining Record, 290 Fillmore Street, Denver, Colorado 80206. Weekly, $12 year. Publishes dependable news reports on mining operations, stock market, etc.

Engineering & Mining Journal, P.O. Box 515, Hightstown, New Jersey 08520 Monthly, $10 per year. Publishes world-wide mining news and highly technical articles.

California Mining Journal, Drawer 628, Santa Cruz, California 95061. Monthly, $8 per year.

Some of the weekly papers in the county seat towns in mining districts publish some interesting and downright educational columns. They carry the local news and occasionally publish some interesting articles on the processes and inventions of local or area miners.

COSTS & PRICES

As you read through any of these Placer Miner's Manuals, you will note that there is a wide disparity between the prices you read about and the prices you pay today. These books were prepared and published in the mid-1930s, at the depth of the Great Depression and today we have a government that is out of control and prices that are spiralling upward so fast that it is hard to know from one day to the next what the price of a product is.

At that time, gold mines around the world were closing and the price of gold was approximately $35.00 per Troy ounce. Today, as I write this, the world price of gold is $627.21, down $21.00 from yesterday. Tomorrow, it may be up $25.00 or down the same amount, depending on how the price is manipulated.

Strangely enough, despite the critical U.S. shortage of gold, the government seems to be doing everything in its power to discourage or prevent mining by the small operator. Rangers, agents, and enforcers seem to be combing the hills like ants trying to locate mines that are not complying with downright idiotic laws and edicts of the bureaucrats. Unbelievable as may sound, and unfortunately, most of these enforcers are in no way other than a formal education qualified for their jobs. The criterion for these jobs is a degree in one of the earth sciences but even then with a few connections a flunky with a degree is business adminstration or some other remote science can get a job as an enforcer. It would appear that these enforcers should either have years of experience in mining or else be a living, accomplished gunslinger. There are some sensible inspectors and enforcers who have their heads screwed on right, but the majority are inexperienced, incompetent, and generally unqualified. As if this wasn't bad enough, there might be agents of 3 to 5 agencies snooping around in the same area and the miner can become totally disoriented and thoroughly chagrined by their rules and edicts. Some of their rules and edicts directly contradict those of the others, and it is the poor miner and taxpayer who is caught in the middle. Yes, indeed, the public servant has become an enforcer and I know that hundreds of experienced miners will support me in this opinion.

These eager-beavers can double or treble your cost of extracting the gold, or silver, from the earth and they could hardly care less. It is a pitiful state of affairs.

SELLING YOUR GOLD

For about 25 years, I have counselled all miners to hoard their gold and all silver and sell only when they have to do this to meet expenses. It has been relatively easy for weekend prospectors and casual miners to do just this and the wisdom of my counsel was proven in early 1980. Many of the wiser miners segregated their gold and sold all but the nuggests and larger flakes at the peak. Some then turned right around and bought volumes of gold when the price dropped abruptly.

I want to wire you in on some gold selling 'secrets' that I doubt if you have ever read in any other book or article. Many, many businessmen feel that their businesses are destined for bankrupcy and they are dedicated to salvage what they can from a lifetime of work in their businesses. A surprising number of them have the knowledge and facilities to test, weigh, and store gold with almost unerring accuracy. Most of these businessmen will buy your raw placer gold at the spot price of the day before or the day they buy. In other words, if you take 10 ounces Troy of 870-fine placer gold to them on a certain day and the spot price is, for example, $600, they would pay you $6,000 for the 10 ounces Troy without deducting for the fineness. That is, they would not offer to pay you 87% of the $6,000. Some buyers will insist on paying only the actual market value of the gold, but quite often when you start to walk out, they will come across with full payment. This is, of course, getting a premium price for your gold.

What actually happens is that these businessmen (as well as doctors, lawyers, bankers, contractors, farmers, ranchers, service station operators, and others) are syphoning money or profits from their businesses to buy the gold from you. Lawyers and bankers may buy in their own interest or in the interest of others. These funds can easily be covered by such pretexts as the service stations shrinkage of gasoline from the tanks, or shoplifting losses from the store, or unreported fees by the lawyer, or remodeling jobs that are not entered into the books of the contractor, and so on. One banker that I know about buys gold with every cent of the notary fees that he collects and it is a surprising amount. I have heard of a Kansas farmer who trades wheat for gold. The wheat is hauled to a chicken farm owned by the miner. The chickens and the eggs are sold to stores over a wide area and, as a consequence, the no part of these transactions shows up on the books of any of the participants.

For a number of years, a small shop owner in southern California has operated a specialty shop in a deluxe shopping center. Most of his business is credit card business but if you pay cash you get a 10% discount and then you get another 5% for not using your credit card. Cash sales are recorded on a cash register at the back entrace of the stop and credit card sales are rung-up on a different cash register at the front of the store. This is in an affluent area where some of the people understand the value of money and discounts. The newly rich and those who do not understand the value of money keep the credit card register hopping while the old and established wealthy people like to take advantage of the long dollar. It works out beautifully for this merchant because he can occasionally give healthy discouts to the credit card customers to keep his mark-up ratio within perspective for the tax birddogs and his cash sales can be charged off to shoplifting, damage, and end-of-season sale losses. Over the years, this merchant has amassed a fortune in gold and it can safely be assumed that his hoard, at $600 per ounce Troy, is worth well into 6 figures.

These practices are, of course, unlawful and have awesome penalties if the perpetrator is caught. Many of these people assume that if corruption from the White House on down to the local city or county clerk is the style, they might as well line their pockets, too. In fact, the integrity of practically any congressman is so low that public opinion polls rate them below the Nixon administration.

If you have gold to sell, your best bet for premium prices to start with is to weight your gold out in 10, 25, or 100 ounce Troy lots. Rember that word TROY because that is the unit used to buy and sell gold. More about this later. Dress neatly and cleanly, even if you have to wear overalls, and start calling on some of the businessmen in your area. Be sure you talk to 'the man' himself and not a clerk or somebody else. Make it a point to call him aside or go to his office and then ask him if he is interested in buying some gold. If he is not, and this is usually the case on the first approach, ask him if you would like to leave your name and address for his use in the event that either he or some of his friends become interested. A business card such as the 'I Find Lost Money, Property, & Jewelry' cards that are available to treasure hunters from the Prospector's Gazette are a dandy entee because they generate interest and curiosity.

Some prospectors and miners who have used this system have become discouraged and sold their gold to lukewarm (?) buyers at bargain basement prices, only to receive several calls weeks or even months later that would have been highly profitable.

It is possible, and quite likely, that once you get a few buyers lined up and they learn to trust you, you will be able to sell more gold than you can produce at your mine. In this perplexing event, you should prepare beforehand to buy or broker the gold from other miners whose property is in your area. It is possible that they already have buyers lined up, but this is unlikely. Many miners, in fact most of them, do not know how to sell their gold to get premium prices and this is where the man (or woman) with the contacts and a reputation for unquestioned honesty can move in and help the miner dispose of his gold and make a few extra bucks himself. You will find that with experience you can make a pretty close estimate of the purity of the gold you handle or mine, but you will still need a test kit, eventually. These kits are usually a small wooden or plastic box that contains 2 acid bottles and a set of test needles, plus a streak plate. To make a test, you merely draw a line on the plate with your gold specimen and then on each side of it, you draw similar lines with the test needles that you estimate approximates the purity of the gold specimen. The lines are then lightly coated with acid and if the reaction of the specimen lines is the same as with one of the test needle lines, you can safely assume that the specimen is of the same purity as the test needle. The instructions that go with the kit are far more definite than my inadequate instructions and I need not waste time and space dwelling on this.

I mentioned before that Troy weight is the universal standard for buying and selling gold. This is not always the case anymore because the inexperienced buyer or seller can get swindled good if he does not understand the differences between Troy and Avoirdupois weights. Consequently, for this reason and others, more and more buyers and sellers are using metric weight to measure their gold deals.

An ounce Troy consists of 31.103 grams and an ounce avoirdupois consists of 28.35 grams. On the other hand, a pound Troy consists of 373.24 grams while a pound Avoirdupois consists of 453.592 grams. Thus, and this has happened far more often than you might believe, it is possible to buy using one of these weights and sell using the other and lose your shirt.

I have included in this book the Gold/Silver Weight Conversion Table that is being used more and more in run-of-the-mill gold deals. This table has saved small fortunes for some buyers and sellers and it has saved leseer amounts of money for others. One of the chief benefits is that you can use any accurate scale to weigh your gold and still be close to laboratory scale accuracy. I have seen gold weighed on the postal scale of the lobby in the Santa Ana, California postoffice and on the scale at the little store in Boron, California with reasonable accuracy. It is not necessary to spend

$150 to $500 for an accurate gold scale. A $25.00 handloaders powder scale is plenty accurate enough and if you want to get almost laboratory scale accuracy you can buy an Ohaus 2610g Triple-beam Balance at almost any sporting goods store for around $75.00. This scale weighs in grams and it is a simple matter to convert the gram weight to Troy or to calculate the price per gram by dividing the spot price for gold (or silver) by 31.103 and buying by the gram instead of by the ounce Troy.

This table, I believe, is accurate to 3 decimal places, or 0.001 but you should check every transaction to be sure, and you should always ask your buyer or seller, whichever he may be, to double check your figures. It is extremely easy to use once you understand the table and it is actually simplicity itself.

Like the 'experts' who graduate from our colleges and find that what they learned in school has to be supported by experience, assuming you are a rank beginner, I will tell you that you are not going to learn all about mining, prospecting, or dabbling in the gold market by reading a book, irregardless of how authentic and detailed the book is. You will be wise to talk with and learn from old experienced miners. You can even be deceived by appearanced and hot-air until you get some mining experience by the handle. There are a number of old fogeys running around with beards, worn clodhoppers, and all of the other TV resemblances of old prospectors that have neither the proverbial pot nor a window to throw it out of. Today's prospector does not look like the movie or TV version so don't be misled. If you will talk to some of them and listen closely, you will find that it is totally impossible for them to have packed all of their experience into one lifetime and you will probably assume that the timing chain between their ears has slipped a cog or two. You could be right, too.

METAL DETECTORS

'Nugget-shooting' is a term I applied to prospecting for gold back around 1952 or 1953. I owned a Fisher T-10 metal detector that I used out on the southern California deserts. My pardner, Hardrock Hammond, had a Detectron and we would often have contests to see who was the best with the different detectors. It was not until the mid-1960s that I found anybody who was a real whiz at nuggetshooting and this was Montana Larry Campbell. He got on a plateau in Arizona and practically got rich sniping coarse gold with his metal detector where others had achieved appreciable success by raking the soil and then 'eyeballing' for the larger particles.

When I got into suction dredging, I finally realized that I was foolish to work the entire bed of a stream, like everybody else was doing, when it was probably possible to detect the stringers of gold and black sand with a metal detector. My theory was that it would be easy to use a waterproof metal detector to scan the bed and banks of a stream and then mark the hot-spots with markers fashioned out of welding rod with a bright-colored ribbon at the top. The idea was nearly perfect under some conditions. The exception to the rule was that some streambeds and some areas of streambeds were so riddled with mineralized conglomerate that the metal detector was a waste of time and effort.

Even in dry placering, the metal detector can often be a valuable tool that can steer you to the richest areas. Here again, it is going to take a lot of experience so that you can understand what your detector is telling you. It is important for the metal detector operator to know that he will rarely run into areas that will indicate a heavy concentration of gold. What you are really looking for is a heavy concentration of black sand. This is not an infallible rule, but you will generally find that if the black sand is not there, the gold will not be there, either. This is true with either wet or dry placer ground.

With respect to lode mines, Hardrock and I were pardners in several lode mines in southern California and Arizona. One of our mines near Essex, California was a lease deal and according to reports an awful lot of gold had been taken out of this mine prior to World War II. It has been shut down because the equipment was needed for the war effort. We thought we were making history when we started to use our metal detectors to locate the rich veins that we had heard about and could not find. We did not find a thing and gave up because we thought our detectors were infallible

A few years later,around 1963, as I recall, I bought a Fisher beat-frequency metal locator. There are those who claim, by the way, that Fisher never made a beat-frequency detector but I still have one that is a collector's item and still in like-new condition. I was working in Rochester, Minnesota on an engineering job and could not get away, so I sent this locator out to Hardrock to test and use. He was quite excited about this locator because it would do things that the others wouldn't. Finally, he took it out to the Essex mine and found vein after vein in the mine that the other detectors had found 'barren'. In case you wonder, we had let our lease lapse and the owner, who accompanied Hammond on the survey, decided to operate the mine himself.

By mid-1964, I had moved down to Weeping Water, Nebraska where I opened the Exanimo Establishment. One day, Herman Aaronson of Weeping Water and I were out to an old stage and freighting station with our infernal machines and I got a good signal with the Fisher beat-frequency locator. Herman tried his transmitter-receiver detector on the same spot. We dug the place out and recovered a jar of old coins. Later, when we were down in a narrow cellar on a farmsite near Louisville, Nebraska we had the same thing occur. It was then that I discerned that a transmitter-receiver detector would not operate properly underground or in narrow confined places. Later, in Colorado and New Mexico, I found this was true and I also found that the transmitter-receiver type had inhibited sensitivy in deep arroyos and in adobe houses.

When I corresponded with various detector manufacturers, it was no surprise that the manufacturers of transmitter-receiver types were horrified at my findings and the manufacturers of beat-frequency types were ecstatic with my findings.

Comes now the VLF (very low frequency) type detectors within the past 6 years and there are those that claim that the VLF-type will work anywhere. About this, I cannot say because I have not had a chance to test any of them under these working conditions.

A metal detector can be a very valuable tool for the prospector and miner but it still has some shortcomings. For one thing, I have heard experts say that they can detect any kind of ore with their detectors, yet I have never seen nor heard of any detector that will detect unerringly certain types of silver ore, and silver is big casino today. For example, I have some very rich silver ore from the Leadville, Colorado area and none of my detectors give any more signal on this ore than they will on a bedbug. I took a 5-pound sample of this ore to the big treasure show at Shepherd Mall in Oklahoma City one year and there was not a detector in the place that gave a signal on this ore. In fact, one of the artifial would-be greats in the detector field tested the sample with several detectors and assayed it as a piece of 'false bedrock'

I have seen rich cinnabar ore that would give a big signal and from the same pit I have seen ore that was simply oozing with droplets of mercury and the latter sample would not get a single squeak from a detector. If I could get more experience in the field, I might derive the cause and the effct of this phenomena.

I have been using the following detectors in recent years with suitable results: a Spartan 175, which is not made anymore; A garrett ADS, a Metrotech 220 and a Metrotech 330. I consider these among the best detectors ever manufacturered commercially, but each has its peculiar shortcomings. Any metal detector is a compromise of electronic theory, at best, and you have to give something to get something. Nevertheless, they are a valuable tool for the prospector.

EQUIPMENT

One of the things that continually bewilders me is this national mania for uniformity, conformity, and follow-the-leader. On my travels around the country, I never cease to be amazed at the people who are participating in this jogging craze and who have to wear the latest jogging shoes, a sweat outfit, and I guess it is some sort of maternity belt that some of these klowns are wearing. I see in some of the medical columns in our newspapers that a good, long, brisk walk is every bit as beneficial as jogging, but if you would tell one of these huffing, puffing bundles-of-suet this, they would probably poleaxe you because walking would not justify the expensive, special garb and they would not attract as much attention.

So it goes with prospecting and mining outfits. You don't need a Coleman stove, nor a gas lantern, nor a motorhome, nor a lightplant, nor a camp table and chairs, nor 90% of the other garbage and junk that is widely advertised to the gullible public. Some of the best meals I have ever eaten in my life were enjoyed with my fanny on the bare ground and my back against a boulder or a tree stump. The best coffee I have ever enjoyed anywhere came right out of a 3-gallon boiled coffeepot, and the best meals I have ever eaten were cooked entirely and completely on a castiron griddle over a common campfire.

Gardner and Johnson have done very well in outlining and listing the provisions and equipment that is needed for prospecting. The only exception I can possibly take with them is their omission of a griddle from their list.

A griddle is one of the most versatile utensiles because an entire meal for 2 to 6 people can easily be cooked at one time on a griddle. This versatility is limited only by your imagination. Just for example, you can fry your bacon in the morning and slide it over to the side to keep warm while you fry your potatoes. Slide them over to the other side and fry your eggs, and when they are done, dish the whole shebang up in individual servings and that is all there is to it.

They tell you and show you, among other things, how to build various pieces of equipment to get started and the chances are that you will build better hardware than you can buy for $200 to $300. I designed the Spartan dredge and this is a reasonably intricate piece of equipment and I do not recommend that anybody other than an experienced mechanic should try to build one of them even though the complete plans are in the Gold Dredger's Handbook. On the other hand, Hammond and I designed and built a number of air-blown dry concentrators back in the 1950s that would be easy for a home craftsman to build. I hope to have these plans in a low-cost book for the prospector eventually.

The 'Tips for Prospectors'column in the Prospector's Gazette gives dozens of tips, suggestions, and ideas for a better camp life and most of these are not to be found in these books by Gardner and Johnson. These tips tell the reader how to save time and money and avoid trouble in their everyday camp and prospecting routine.

STATE SERVICES & PUBLICATIONS

Every state has a mineral, mining, or geology division of some sort or another. Some of them hardly justify the name while others are very good and provide a variety of services and publications for the benefit of the prospector and miner. Some of them provide assay and mineral identification for residents of the state and some will even provide inspection services. I am not going to list the services that are provided because these bureaus and departments operate at the whim of the state legislatures and it is not uncommon to have some services curtailed while others are expanded, and then have them reinstated again or completely discontinued.

Here is a list of the better state agencies with their addresses. If you are seriously interested in or involved in prospecting and mine, I recommend that you write to each of them and get a pricelist of their publications because some of them are very good and all of them are helpful.

Arizona Bureau of Mines, University of Arizona, Tucson, Arizona 85721. You will find their publications brief, concise, and very helpful, and they cost very little, too.

California Division of Mines & Geology, Box 2980, Sacramento, California 95814. They have some superb books and pamphlets dealing with mining in California and they provide many helpful services to the prospector and the miner.

Idaho Bureau of Mines & Geology, Moscow, Idaho 83843. They have some fine practical mining publications at quite reasonable prices. Some of them are profusely illustrated and have been used widely by miners across the country for years.

Montana Bureau of Mines & Geology, Montana College of Mineral Science and Technology, Butte, Montana 59701. Their publications are very good and at one time they distributed some very good mimeographed bulletins listing mining locations throughout the state.

Nevada Bureau of Mines, University of Nevada, Reno, Nevada 89507. Their publication is very good and they did provide a variety of services to state residents.

New Mexico Bureau of Mines & Mineral Resources, Campus Station, Socorro, New Mexico 87801. They have a variety of technical publications that even include tour maps for all sections of the state that describe the geology.

Oregon Department of Geology and Mineral Industries, 1069 State Office Building, Portland, Oregon 97201. I have not seen their publications list for years but they always maintained a very helpful list of publications at sensible prices.

Washington Division of Mines & Geology, P.O. Box 168, Olympia, Washington 98501. Their publications are outstanding and at one time they had memiographed sheets showing the principal mining areas throughout the state. In fact, I was surprised to find a mention of suction dredging dating back to the turn of the century in one or two of their publications.

GETTING PAID TO DO IT!

You can get federal assistance for your prospecting and mining ventures from the U.S. Government, but unless you are a combination attorney, accountant, geologist, and mining engineer, I advise you to steer clear of this circle of confusion. Half of your time will be spent keeping books and records, making reports, and waltzing inspectors around the place. If you are a fearless personality who enjoys paperwork and company, you can write to: The Office of Mineral Exploration, U. S. Geological Survey, Washington, D.C. 20252 and inquire about their requirements and procedures. They will refer you to one of their four regional offices and you can take it from there.

HEADS UP

Some modern prospectors and miners have become wealthy by using diligence and common sense in their prospecting and mining. The right books will tell you how, and even where, but it takes good judgement and plenty of experience to assure success. You have to observe and study to do it and it is unlikely that you'll make it if you spend your sparetime watching the boobtube or if you waste your time on other pleasurable activities. You have to invest time as well as money and the investment of time and effort is far more important than the cash investment.

In fact, I know several very successful gold miners who were eating beans without the bacon when they struck it rich, and I have never heard one of them crying about the restricted diet or the hard times before they hit.

If you do start hitting the big payoff, keep building up your reserves and don't rush to town to let everybody else know about it, or you are liable to wind up on the outside looking in, as has happened so many times. There is one book that I would like for everybody to read. It is RICHEST MAN IN BABYLON, by George B. Clason. You can get a paperback copy for about $3.00 plus postage at most bookstores, and if they can't supply you, Exanimo Establishment, Box 448, Fremont, Nebraska 68025 can supply you. This book tells the secret of acquiring wealth, and anybody can do it. This book might well be mandatory reading for every high school student. You do as you wish, the least I can do is recommend the book to you, and if you are unable to follow the easy, simple process, your mirror will show you who is at fault.

Your public library has a wealth of helpful information waiting for you. If you do not use the services available, somebody else will. Libraries have an interchange system so that practically any book is available to you. The librarian will be glad to explain the indexing system to you. If you can learn it in college, you can also learn it in the library.

I have this book and the other two Placer Miner's Manuals help you find a big bonanza, or even a small one will make you wealthy, so that you can enjoy life the way it was meant to be relished. Thousands have done it, and you can, too.

Kindest regards, best wishes, and

Exanimo!

Karl von Mueller
Exanimo Hacienda
Segundo, Colorado 81070

I. C. 6786 September 1934

Department of the Interior

United States Bureau of Mines
John W. Finch, Director

INFORMATION CIRCULAR

PLACER MINING IN THE WESTERN UNITED STATES

PART I. GENERAL INFORMATION, HAND-SHOVELING, AND GROUND-SLUICING

BY

E. D. Gardner and C. H. Johnson

I. C. 6786,
September 1934

INFORMATION CIRCULAR

UNITED STATES BUREAU OF MINES

PLACER MINING IN THE WESTERN UNITED STATES[1]

Part I. - General Information, Hand-Shoveling, and Ground- Sluicing

By E. D. Gardner[2] and C. H. Johnson[3]

CONTENTS

1 The Bureau of Mines will welcome reprinting of this paper, provided the following footnote acknowledgment is used: "Reprinted from U.S. Bureau of Mines Information Circular 6786."

2 Supervising engineer, U.S. Bureau of Mines, Southwest Experiment Station, Tucson, Ariz.

3 Assistant mining engineer, U.S. Bureau of Mines, Southwest Experiment Station, Tucson, Ariz.

INTRODUCTION

Placer mining is the mining and treatment of alluvial deposits for the recovery of their valuable minerals. The method has been used principally for mining gold, but a large proportion of the world's production of tin, platinum, and diamonds and other gem stones and minor quantities of other heavy minerals have been won in this manner. In the United States, as in the world at large, gold has been the principal mineral obtained by placer mining. Minor quantities of metals of the platinum group are recovered with the gold in some localities. Important quantities of sapphires have been produced at placer mines in Montana, and tungsten minerals have been obtained on a commercial scale from placer deposits in California and Colorado. Other heavy minerals or gem stones, however, have not been mined to any important extent by this method in the United States.

The search for placer gold and the working of the deposits when found have had much to do with the early development of the West. Placer mining has been gradually overtaken and surpassed in importance by lode-gold mining, until in 1932 less than a quarter of the country's total gold production was from placers or about an eighth, excluding Alaska. In 1932 about 76 percent of the placer gold produced in the United States was recovered by dredging. Although other forms of placer mining still are important, they have been declining for many years, as the richest and most readily mined deposits of gravel along the stream courses have been exhausted. During 1931 and 1932 there was a revival of small-scale mining, but few new deposits were discovered.

California has ranked first in the production of placer gold since the discovery of gold on Sutter Creek in 1848. In 1932 the relative importance of the other Western States in gold production by placer mining was as follows: Oregon, Idaho, Nevada, Montana, Arizona, Colorado, New Mexico, South Dakota, Washington, Utah, and Wyoming.

This paper deals with the history of placer mining and production of placer gold, geology of placer deposits, location of placer claims on public lands, sampling and estimation of gold placers, and the classification of placer-mining methods, together with discussions of hand mining and ground sluicing.

Two subsequent papers[4] deal with other phases of placer mining. All phases of placer mining are discussed in the three papers and current practices are illustrated in descriptions of individual mines.

ACKNOWLEDGMENTS

The authors have drawn freely upon the available literature on placer mining, geology, engineering, and other allied subjects; they have endeavored to make suitable reference throughout the text.

G. A. Bigelow of San Francisco furnished data on methods and costs of sampling placer ground. Victor C. Heikes, Clarence N. Gerry, and Chas. W. Henderson of the Economics Branch of the Bureau of Mines kindly provided the authors with lists of the producers of placer gold; this information facilitated field investigations.

The authors also wish to acknowledge their indebtedness to the operators of placers in the Western States who generously supplied information without which this paper could not have been written.

---

4 Gardner, E. D., and Johnson, C. H., Placer Mining in the Western United States: Part II. - Hydraulicking, Treatment of Placer Concentrates and Marketing of Gold, and Part III. - Dredging and Other Forms of Mechanical Handling of Gravel, and Drift Mining: Inf. Circs. 6787 and 6788, Bureau of Mines, 1934.

HISTORY OF PLACER MINING AND PRODUCTION OF PLACER GOLD IN THE WESTERN UNITED STATES[5]

The earliest large placer-gold production in what is now the western United States was from the Old and New Placer diggings, near Golden, Santa Fe County, N. Mex., which were worked as early as 1828 and have yielded probably 3 or 4 million dollars in gold. Several smaller deposits were likewise known and worked, but the first discovery of major importance was made by James W. Marshall on January 24, 1848 [6] Marshall was building a sawmill on the south branch of the American River at Coloma, 40 miles west of the present site of Sacramento, and found gold nuggets in the tailrace of the mill. The news spread too slowly to benefit Mexico, which on February 2, 1848, upon payment of $15,000,000, ceded not only California, but nearly all the present Southwest to the United States as war indemnity. The rapid settlement of the West began with the great gold rush to the California fields.

The excitement over gold spread through central California in 1848. In that and the next year the placers of the Trinity and Klamath Rivers, in northern California, were discovered. The total California production of these 2 years was $40,000,000.

Oregon settlers, moving to the southern gold fields, found rich placers in the Grants Pass district in 1852. A brief pause ensued, but when gold was found on the Fraser River and the Caribou, in British Columbia, in 1858 and 1860, the consequent stampedes rapidly opened northern and central Washington.

In 1860, E. D. Pierce discovered gold in the Clearwater country, now part of Idaho. Within a year Pierce City, Oro Fino, Elk City, Florence, and Warren, in Idaho, were founded; gold was also discovered during this period on the John Day and Powder Rivers in eastern Oregon. The Boise Basin, Idaho, was entered and rich placers found, in 1862; others were discovered at Silver City, Idaho, in 1863. The Gold Creek, Bannack, and Alder Creek mines were discovered in Montana in 1862.

The Territory of Idaho was created early in 1863, with Boise as its capitol, and Helena, another placer camp, became the capitol of the Territory of Montana, organized in 1864.

The Gila City placers in Arizona were found in 1858 and the greater ones of the La Paz district in 1862. A year later the Weaver and Lynx Creek deposits were booming.[7]

In Colorado a placer-mining expedition found some gold on Cherry Creek, Ralston Creek, and the Platte River in 1858 but nothing of importance until the next year. The change from the first skepticism of 1848 to the Nation-wide gold fever of the following period is shown interestingly by the fact that the mere presence of prospectors in Colorado gave rise to the wild rumors of rich diggings "near Pikes Peak", and brought a large population to the winter camp on the Platte, where Denver was founded, before a single valuable deposit was discovered.[8] In 1859, rich placers were located on Clear Creek, in the South Park, and on the upper Blue and Arkansas Rivers. Colorado was made a territory in 1861, with the seat of Government at Denver.

Nevada's first placer mining was done by settlers in Carson Valley whose chief livelihood was in supplying the California wagon trains. From 1849 on, sporadic placering was done in Gold Canyon on the side of Mount Davidson. Gold was found in a near-by canyon, Six-Mile Creek, in 1857 and 2 years later placer workings there uncovered the outcrop of the world-famous Comstock lode. Further placer discoveries took place in the Sierra district soon after 1863, in the Tuscarora district in 1867, at Copper Mountain (Charleston district) and Osceola in 1876 and 1877, and at Spring Valley in 1881.[9] Other discoveries of some import-

---

5 Chiefly from Bancroft's works, except as otherwise acknowledged.

6 Rickard, T. A., The Discovery of Gold in California: Univ. of California Chronicle, vol. 30, no. 2, April 1928, pp. 141-169.

7 Wilson, E. D., Arizona Gold Placers and Placering: Univ. of Arizona, Arizona Bureau of Mines, Bull. 135, Aug. 15, 1933, pp. 13, 14.

8 Henderson, C. W., Mining in Colorado: U. S. Geol. Survey Prof. Paper 138, 1926, pp. 1-8.

9 See Smith, A. M., and Vanderburg, Wm. O., Placer Mining in Nevada: Univ. of Nevada Bull., vol. 26, no. 8, Dec. 16, 1932.

ance have occurred even more recently in this State, as for instance, at Round Mountain in 1906; the production has been about $1,300,000 since its discovery. Lack of water may be partly responsible for the fact that many of Nevada's placers were not prospected until long after the heyday of placering in Idaho, Oregon, and California. The most productive placers in the State, those at Spring Valley, were not worked until 1881, although the lodes there were mined as early as 1868.

In Utah placer gold has been of little importance. The largest production from any district, about $1,500,000, came from Bingham Canyon, where lode discoveries in 1863 were followed a year later by placer locations.[10]

Placer gold was discovered in South Dakota in July 1874[11] by Horatio Ross, a prospector who had accompanied General Custer's reconnaissance expedition into the Black Hills. In 1875, as a result of the inevitable stampede, the rich placers of Deadwood Gulch were found, leading shortly to the location of the Homestake lode mine.

The only New Mexico placers credited with a large production, other than those of the Golden district already mentioned, were discovered in the Elizabethtown district in 1867 These were rich enough to encourage extensive hydraulic installations, including a ditch 41 miles long.[12]

The principal yield of gold from most districts has been in hand methods during the first few seasons following discovery. Hydraulicking has prolonged the period of activity in many districts, but in only a few instances has resulted in greater yields than handworking in the first few years. Regular production nearly comparable to lode mining has been established by dredging in a few fields. This sequence of phases of placer mining is shown by the history of the Boise Basin district in Idaho. Gold was discovered there in 1862; and, according to one estimate, $3,000,000 worth of gold was taken out in 1863, $4,000,000 in 1864, and $5,000,000 in both 1865 and 1866.[13] Hydraulicking was introduced in 1867; the yield, however, dwindled nearly 20 percent annually for the next 10 years and more slowly thereafter, until it was only $200,000 in 1898. Dredging which began late in 1898 doubled the annual production the first year. Again after 1910, when the placer yield was only about $130,000, a new dredging project boosted production to a peak of nearly $500,000, although hydraulicking continued to dwindle. There was virtually no dredging from 1916 through 1925; in 1920 only $1,600 worth of placer gold was mined. A dredge started up again in 1926, and for 5 years, until the boat burned in 1930, the annual yield was $40,000 to $60,000.

The Manhattan district, in Lyon County, Nev., is an example on a small scale of the trend of production in most placer camps, except that here growth was relatively slow due to most of the gold being produced from drift mines. The first recorded placer production was in 1908, when $16,000 was produced. The peak was reached in 1912 and 1913 when each year was credited with a yield of about $165,000. Since then production has fallen off an average of about 20 percent each year, until in 1931 it was only $1,400.

---

10 Butler, B. S., The Ore Deposits of Utah: U.S. Geol. Survey Prof. Paper 111, 1920, p. 340.

11 Lincoln, F. C., Half A Century of Mining in the Black Hills: Eng. and Min. Jour., vol. 122, Aug. 7, 1926, pp. 205-206.

Also O'Harra, C. C., Early Placer Gold Mining in the Black Hills: The Black Hills Engineer, South Dakota School of Mines Quarterly, vol. 19, 1931, pp. 343-361.

12 Lindgren, Waldemar, and others, Ore Deposits of New Mexico: U.S. Geol. Survey Prof. Paper 68, 1910, pp. 92-105.

13 U.S. Geol. Survey Mineral Resources of the United States, 1914, part I, p. 616.

History of Dredging

The story of dredging in the United States begins in Montana, where the first successful bucket dredge was built by the Bucyrus Co. on Grasshopper Creek, Beaverhead County, in 1895. This boat was rebuilt at the end of 1895 and 1896 seasons, and in 1897 two new dredges were put in commission at the same place. In 1897 the first California bucket dredge was designed by R. H. Postlethwaite and built by the Risdon Iron Works on the Yuba River. The next year the Risdon Co. likewise built two dredges on the Blue River near Breckenridge, Colo., and one near Idaho City, Idaho. In 1899 four more dredges were built in the latter district. The Yuba River dredge soon was wrecked owing to the swiftness of the river, but in 1898 W. P. Hammon and Thomas Couch put another Risdon-built bucket-elevator dredge in operation on the Feather River at Oroville. This, like the Postlethwaite dredge, was of the single-lift stacker type contrasted with the double- and later single-lift flume-type dredges built by the Bucyrus Co. In California the single-lift stacker type was adopted almost universally, although trials were made in 1899 and 1900 of the double-lift dredge. In Oregon, Idaho, and Montana the flume-type dredge persisted to some extent.

The first successful bucket-type dredges in this country were built according to the New Zealand practice. At the outset, however, these boats were far from satisfactory, and 2 or 3 years of experimentation were necessary before it was known that dredging could be done profitably under the generally severe conditions encountered. After the best general type of dredge to use was known the industry grew rapidly. By 1910 a hundred dredges had been built in California alone, representing an investment of about $9,000,000. Of these, 63 were operating in that year. Dredging has renewed or prolonged the life of many placer districts for years. It represents usually an end phase of placer mining, never the pioneering activity which has built new communities and even States. Historically its chief importance lies in the constant production, year after year, of 4 to 8 million dollars in gold from the huge Californian fields, and 1 to 4 million from Alaska.

The total dredge-gold production of California[14] from 1898 through 1932 has been about $175,000,000. Of this about $67,000,000 has come from the Yuba River field near Marysville since 1903. The American River field around Natoma, 10 to 20 miles above Sacramento, is the next most important field, having yielded about $45,000,000 since 1900. The dredge production of gold in the Feather River field in Butte County, where the first successful dredging was done in California in 1898, has been about $34,000,000.

In recent years (1925-32) about 25 boats have operated in California, producing annually about $4,500,000 from the handling of 45 to 50 million yards of gravel.

The value of the silver recovered by Californian dredges is small compared to that of the gold, being less than 0.2 percent of the latter in recent years. The relative quantity of the two metals recovered annually since 1920 has ranged from 10 to 17 ounces (about 15 at present) of gold per ounce of silver.

Of the placer-gold production of Alaska in 1932 about 80 percent, or $4,300,000, was mined by 25 dredges. In 1931, 13 boats in the Yukon region produced about $2,600,000; 14 in the Seward Peninsula and one in the Kuskokwim region produced the balance, about $1,100,000. The total yardage dredged in 1931 is estimated at 10,200,000 yards, yielding an average of 37 cents per yard.[15]

---

14 Winston, W. B., and Janin, Charles, Gold Dredging in California: California State Min. Bur., Bull. 57, 1910.
Also, Mineral Resources of the United States, U.S. Geol. Survey and U.S. Bureau of Mines, for the years concerned.

15 Smith, P. S., Mineral Industry of Alaska in 1931: U.S. Geol. Survey Bull. 844, 1933, p. 52.

Production

Table 1 shows United States dredge production of placer gold and the number of dredges operating, by States, from 1896 to 1932.

Table 2 shows the production of placer gold in the United States, by States, to the end of 1932. The total estimated production of each of the States before 1901 is given in the table, but many of these data are based on scanty evidence. Since 1901 the production is given by years for each of the States in which placer gold was produced. It will be noted that the percentage of total gold produced in the United States by placering increased from 20.4 in 1931 to 23.4 in 1932. This was due mainly to the great increase in small-scale operations throughout the country. In the Western States in which no dredging was done, the increase of placer gold was several fold.

GENERAL GEOLOGY OF PLACER-GOLD DEPOSITS

Placer-gold deposits result from the weathering and erosion of gold-bearing rocks. Change of temperature, water, and natural solvents disintegrate the rock and partly free the gold from its gangue. Running water transports the products of weathering seaward, meanwhile grinding them to smaller and smaller sizes and thus setting free more gold. Because of its high specific gravity the gold settles through the moving mass of silt, sand, and gravel being carried by streams or floods, and most of it is left behind as the lighter material is carried onward. It comes to rest when the velocity of the stream is insufficient to carry it farther and usually becomes concentrated on bedrock.

The formation of rich placers is favored by the peneplanation or baseleveling of an area, which results in very deep disintegration of the rock surface. When this is followed by uplift and renewed rapid erosion, minerals such as gold or platinum in the great masses of rock are concentrated in the stream channels. This has been the geologic history of many important placer districts.

Gold placer deposits generally are found in districts where lode gold deposits occur. Typically the lodes comprise numerous thin quartz veins with small but rich ore shoots; some of the gold is coarse and free-milling, that is, not intimately combined with other metallic minerals. The lode deposits may be too small or too low-grade to be of commercial value; in some places they may have been removed entirely by erosion.

The discovery of gold in a present stream bed is followed logically by searching for bench gravels, that is, remnants of early stream gravels now lying at relatively higher elevations because of the deepening of the stream bed. This point was made by Mertie[16] with reference to interior Alaska but it applies to many other districts, such as the Sierra Nevada, where some of the deposits laid down during an earlier (Tertiary) cycle of erosion have been exceedingly rich and productive.

Placer deposits usually are not found in the extreme upper portions of streams where the gradient is steep, but under exceptional conditions enough coarse gold might concentrate even there to form valuable deposits. Generally extensive placer-gold deposits are formed just below the steeper grades, where the streams emerge into the lower relief of the foothills. In the Idaho (Boise) Basin, for instance, the placers are found in the flat, lower stretches of Boise River and its tributaries where the stream-bed grades range from 25 to 50 feet per mile rather than in the headwaters where the grades increase rapidly to 100 or 200 feet or more per mile.

---

16 Mertie, J. B., The Occurrence of Metalliferous Deposits in the Yukon and Kuskokwim Regions: U.S. Geol. Survey Bull. 739, 1923, p. 163.

 TABLE 1.- Production of gold in the United States, by dredges, and number of dredges producing, by States, 1896-1932

(24) (Rearrangement of tabulation in Mineral Resources, 1914, pt. I, p. 855; 1915-30 data from Mineral Resources for respective years)

Year	Alaska		California		Colorado		Idaho		Montana		Oregon		Other States		Total	
	Value	No.	Value	No.	Value	No.	Value	No	Value	No.	Value	No.	Value	No.	Value	No.
1896			$2,000	1					$42,000	1					$44,000	2
1897			5,000	1			$11,436	1	102,120	4					118,556	6
1898			18,887	3			13,920	1	154,893	4					187,700	8
1899			206,302	8			62,436	5	165,440	5					434,178	18
1900			200,929	16			129,443	6	189,665	5					520,037	27
1901			471,762	22	$6,000	1	116,117	6	146,134	5					740,013	34
1902			867,665	29	10,000	1	101,257	10	318,914	7			[1]$71,686	1	1,369,522	48
1903	$20,000	2	1,475,749	31	15,000	1	86,113	6	229,332	4			[2]89,870	1	1,916,064	45
1904	25,000	3	2,187,038	42	65,594	3	99,110	7	245,700	2			[3]101,275	4	2,723,717	61
1905	40,000	3	3,276,141	50	33,342	3	34,336	3	275,542	5			[4]28,015	4	3,687,376	68
1906	120,000	3	5,098,359	59	48,343	3	38,340	3	397,030	4			[5]19,322	4	5,721,394	76
1907	250,000	4	5,065,437	57	35,235	3	74,438	6	197,141	4	$23,191	2	[6]10,260	3	5,655,702	79
1908	171,000	4	6,536,189	69	141,773	4	77,189	5	402,667	4			[7]24,852	2	7,353,670	88
1909	425,000	14	7,382,950	63	404,636	4	101,704	8	426,649	5	42,667	2			8,783,606	96
1910	800,000	18	7,550,254	72	344,211	6	91,247	6	473,318	7	34,010	6			9,293,040	115
1911	1,500,000	27	7,666,461	65	272,173	4	258,791	7	597,778	8	14,575	3	[8]16,591	5	10,326,369	119
1912	2,200,000	38	7,429,955	65	384,748	3	481,077	8	710,387	6			[9]12,744	4	11,218,911	124
1913	2,200,000	36	8,090,294	63	372,288	4	561,876	6	685,210	5			[10]317,268	2	12,226,936	116
1914	2,350,000	42	7,783,394	60	602,655	5	568,989	4	835,615	5			[11]372,130	4	12,512,783	120
1915	2,330,000	35	7,796,465	58	672,386	5	486,541	7	861,626	5			[12]336,107	4	12,483,125	114
1916	2,679,000	34	7,769,227	60	695,265	6	327,696	4	642,572	5	670,415	3	2,539	1	12,786,714	113
1917	2,500,000	36	8,313,527	55	647,270	6	59,446	4	409,455	3	618,922	4	1,805	1	12,550,425	109
1918	1,425,000	28	7,431,927	48	522,921	6	239,762	5	334,750	3	387,740	3			10,342,100	93
1919	1,360,000	28	7,716,919	46	542,103	6	164,854	5	265,590	3	296,750	3			10,346,216	91
1920	1,129,932	22	6,900,366	40	512,876	5	101,679	3	255,550	3	358,884	4	[13]27,169	1	9,286,456	78
1921	1,582,520	24	7,756,787	35	337,950	3	151,762	3	190,416	1	381,960	4	[14]134,173	2	10,535,568	72
1922	1,767,753	23	4,999,215	35	346,327	4	158,827	3	36,941	1	269,994	4	[15]110,211	1	7,689,268	71
1923	1,848,596	25	6,065,735	29	358,864	4	469,900	4			224,117	3	[16]31,835	1	8,999,047	66
1924	1,563,361	27	4,305,521	27	412,080	4	340,462	2			291,557	3			6,912,981	63
1925	1,572,312	27	4,750,842	25	141,103	4	229,489	2			137,282	2			6,831,028	60
1926	2,291,000	32	4,950,545	23	38,860	2	141,160	3			74,191	2	[17]27,029	1	7,522,785	63
1927	1,740,000	28	5,461,929	25	86,902	1	114,116	3			112,643	2			7,515,590	59
1928	2,185,000	26	4,430,913	24	51,019	1	133,418	3			90,103	2	[18]1,878	1	6,892,331	57
1929	2,932,000	30	3,589,259	25	38,497	1	60,143	3			205,464	3	[19]5,938	1	6,831,301	63
1930	3,912,600	27	3,451,801	24	130,824	1	68,527	3			174,470	5			7,738,222	60
1931	3,749,000	28	3,619,355	22	8,793	1	80,352	3			138,155	3			7,595,655	57
1932	4,293,000	25	3,903,481	22	23,194	1	171,130	5			160,848	4			8,551,653	57
1896 to 1932	50,962,074		174,528,580		8,303,232		6,407,083		9,592,435		[20]5,815,684		[21]634,951		256,244,039	

See page 9 for footnotes.

1 New Mexico. 2 New Mexico. 3 Oregon 3, New Mexico 1. 4 Georgia, North Carolina, Oregon, and New Mexico. 5 Georgia, Oregon. 6 Georgia 2, New Mexico 1. 7 Georgia, Oregon. 8 Nevada 2, South Dakota, Georgia, North Carolina. 9 South Dakota, North Carolina, Georgia, Wyoming. 10 Oregon, Nevada, Alabama, North Carolina. 11 Oregon, Nevada, Alabama, North Carolina. 12 Approximately $335,000 from 2 Oregon dredges. 13 Nevada. 14 $133,020 from 1 dredge in Nevada. 15 Nevada. 16 Nevada. 17 Washington. 18 Washington. 19 Washington. 20 Includes $1,107,746 not segregated from Other States, 1904-15. 21 Excludes $1,107,746 produced by Oregon, 1904-6, 1908, and 1913-15, but credited to Other States.

Coarse gold commonly is found in strata that contain a large percentage of material coarser than sand, as the high specific gravity of gold gives it a resistance to transportation nearer to that of large pebbles than to that of sand. Similarly, fine gold or flour gold is found associated with strata of sand and small pebbles rather than in the layers of silt or clay that mark a low stage of the river.

In prospecting a stream it is reasonable to test the gravels on the convex banks or short sides of the bends and near the heads of the bars. It should be remembered, however, that most of the transporting work of the stream is done during flood times, and the location and shape of the pay streaks in the gravel therefore are fixed during a high stage of the river.

In working placer ground, especially old channels, it is often advantageous to know the direction of flow of the stream that laid down the deposit. This may be indicated by the position of flat pebbles and boulders. These usually are "shingled", that is, lie with a distinct tilt downstream; otherwise the current would have tended to lift or turn them over rather than to press them down onto the stream bed.

Gold placers have been classified by Brooks[17] and more elaborately by Mertie.[18] Gilbert[19] presents a detailed discussion of the eroding and transporting action of natural streams. Although his paper is not concerned with the formation of placers it is helpful in understanding the structure of all stream deposits. The most important type of placer deposit from the standpoint of total production is the simple stream placer in which the gravels of present streams, whether permanent or intermittent, contain gold. The gold lies chiefly along narrow pay streaks within the wider channel or at favorable points in bars which may be exposed at low water. Bars form at points of relatively low velocity, hence are found not only adjacent to the convex bank of the stream at curves but also as transverse bars or crossings where the current swings from one bank to the other between curves in opposite directions and in the straight stretch below any curve. Gold will also be dropped where flood conditions permit deposition of part of the stream's load of sand or gravel.

Streams that are depositing sediment rather than eroding their beds tend to meander so that their banks usually consist of earlier-deposited material similar to their present bars. If erosion of a stream bed is renewed because of a later uplift, or a general lowering of the drainage system, some of the older deposits once forming the banks may be left high above the reach of the stream, which now may be sunk into a rock-walled channel. Even after the stream ceases to erode and again meanders through a gentle, flat-bottomed valley filled with new gravel deposits, remnants of the old bars or "benches" may be found high above water level.

---

17 Collier, A. J., Hess, F. L., Smith, P. S., and Brooks, A. H., The Gold Placers of Parts of Seward Peninsula, Alaska: U.S. Geol. Survey Bull. 328, 1908, p. 115.

18 Mertie, J. B., The Occurrence of Metalliferous Deposits in the Yukon and Kuskokwim Regions: U.S. Geol. Survey Bull. 739, 1923, pp. 160-162.

19 Gilbert, G. K., The Transportation of Debris by Running Water: U.S. Geol. Survey Prof. Paper 86, 1914, pp. 219-233.

TABLE 2.- Placer gold production of United States, by States, before 1901 and 1901-32, by years[1]

Year	Alabama	Alaska	Arizona	California	Colorado	Georgia	Idaho	Maryland	Montana	Nevada	New Mexico
Through 1900	[2]$300,000	[3]$14,315,000	[4]$8,200,000	[5]$1,032,827,480	[7]$21,294,219	[8]$12,000,000	[9]$90,000,000	....	[10]$165,000,000	[11]$25,000,000	[12]$14,000,000
1901	1,385	4,980,000	105,034	3,951,049	87,324	18,047	753,716	0	522,700	33,509	59,721
1902	517	5,887,000	10,274	4,247,602	118,774	21,395	365,767	$765	447,046	15,649	130,481
1903	310	6,010,000	11,742	4,052,761	129,049	25,426	378,853	455	481,447	36,424	114,605
1904	(6)	6,025,000	16,848	4,985,290	193,068	[6]52,000	493,002	(6)	478,565	30,192	149,424
1905	1,034	12,340,000	42,667	5,892,076	99,984	29,995	340,465	0	396,901	8,274	99,335
1906	0	18,607,000	40,502	7,375,925	106,019	17,354	353,481	0	521,815	52,838	26,807
1907	42	16,491,000	44,891	6,840,695	97,219	23,413	356,905	0	348,667	55,275	19,340
1908	945	15,888,000	30,937	8,231,187	184,457	11,201	285,643	0	549,995	79,751	23,198
1909	69	16,252,638	28,648	9,104,433	457,085	16,433	281,727	0	543,372	82,965	22,010
1910	357	11,984,806	25,990	8,888,795	389,828	18,211	242,546	0	575,917	162,371	26,094
1911	0	12,540,000	23,641	8,986,527	319,038	23,738	404,327	0	684,801	210,461	18,714
1912	0	11,990,000	43,046	8,645,663	423,865	6,846	632,029	0	806,419	231,653	16,926
1913	0	10,680,000	30,691	8,836,177	408,007	8,570	694,053	0	801,002	305,442	7,861
1914	500	10,730,000	30,140	9,080,849	642,360	11,043	700,454	0	942,217	377,262	29,152
1915	59	10,480,000	35,248	8,608,617	693,310	15,256	584,890	0	949,248	395,319	9,242
1916	777	11,140,000	14,281	8,575,657	712,924	7,626	449,093	0	723,159	354,313	11,116
1917	0	9,810,000	17,214	9,074,030	661,028	2,811	135,231	0	467,063	292,584	12,179
1918	55	5,900,000	4,234	7,838,779	526,202	4,905	276,410	0	396,232	218,380	3,118
1919	0	4,970,000	4,694	8,033,076	550,562	715	190,752	0	291,430	132,288	4,959
1920	0	3,873,000	4,567	7,060,613	514,588	0	113,814	0	288,946	152,639	2,188
1921	0	4,226,000	12,524	8,154,824	344,640	711	181,600	0	227,161	363,142	8,281
1922	0	4,395,000	11,981	5,499,855	356,403	1,723	183,972	0	71,786	239,842	3,932
1923	114	3,608,500	8,854	6,522,583	364,429	513	498,709	0	40,779	81,485	4,218
1924	0	3,564,000	3,139	4,588,372	418,506	79	358,121	0	27,361	27,369	3,639
1925	0	3,223,000	4,267	5,096,144	150,318	68	262,386	0	39,385	52,435	2,018
1926	0	3,769,000	7,007	5,228,403	46,954	1,088	172,826	0	22,828	59,249	2,687
1927	0	2,982,000	6,257	5,837,313	94,434	1,043	155,459	0	22,325	37,400	5,808
1928	0	3,347,000	6,400	4,850,629	61,406	256	169,336	0	17,884	38,266	1,347
1929	203	4,117,000	5,652	3,870,607	45,850	1,928	85,373	0	12,334	43,762	1,650
1930	450	4,837,000	13,057	3,755,143	138,243	243	82,428	0	14,899	38,438	1,316
1931	407	4,842,000	22,103	4,020,746	21,586	781	107,773	0	39,439	59,602	8,405
1932	0	5,522,000	71,933	4,765,475	51,655	3,720	257,151	0	73,125	111,798	26,259
1901-32 (inc.)	7,224	255,010,944	738,463	210,499,895	9,409,115	327,138	10,548,292	1,220	11,826,248	4,380,377	856,030
Through 1932	307,224	269,325,944	8,938,463	1,243,327,375	30,703,334	12,327,138	100,548,292	1,220	176,826,248	29,380,377	14,856,030

See page 12 for footnotes.

TABLE 2.- Placer gold production of United States, by States, before 1901 and 1901-32, by years - Continued

Year	North Carolina	Oregon	South Carolina	South Dakota	Tennessee	Utah	Virginia	Washington	Wyoming	Total	Percent of total U.S. gold production
Through 1900	[13]$5,000,000	[14]$25,000,000	[15]$1,000,000	[16]$7,000,000	....	[17]$1,000,000	[18]$1,000,000	[19]$5,000,000	[20]$500,000	$1,428,436,699	[21]59.9
1901	18,522	1,422,016	7,917	0	0	0	2,646	102,388	41,344	12,107,318	15.4
1902	16,599	243,886	4,672	0	$145	0	558	62,016	45,230	11,618,376	14.5
1903	9,054	471,020	2,625	0	62	0	0	4,906	8,289	11,737,028	15.9
1904	(6)	349,214	(6)	3,614	(6)	1,354	(6)	9,823	2,231	12,789,625	15.9
1905	10,005	251,619	0	9,163	207	6,656	806	6,439	2,116	19,537,742	22.2
1906	11,906	361,560	270	6,250	1,076	8,613	0	19,209	1,385	27,512,010	29.2
1907	9,834	331,406	925	924	0	9,061	117	21,860	4,045	24,655,619	27.3
1908	17,555	272,593	810	9,942	612	9,110	661	19,478	820	25,616,895	27.2
1909	10,848	221,318	1,445	1,179	625	2,525	876	5,988	1,114	27,035,298	27.3
1910	10,281	170,925	2,076	2,972	500	3,980	90	3,859	654	22,510,252	23.8
1911	5,111	168,274	261	12,073	0	5,634	808	3,999	7,041	23,414,448	24.2
1912	8,752	189,096	419	13,725	0	5,680	0	4,728	766	23,019,613	24.9
1913	6,378	450,628	218	1,393	0	1,920	0	4,144	1,407	22,237,891	25.0
1914	6,707	548,317	449	1,405	0	1,231	0	5,756	1,841	23,109,683	25.3
1915	8,486	482,170	248	1,586	0	958	0	7,160	704	22,272,501	22.7
1916	7,893	872,517	320	2,111	0	1,250	0	8,277	349	22,881,663	25.1
1917	3,979	727,366	164	924	0	112	0	5,868	34	21,210,587	26.3
1918	1,631	498,249	0	431	0	1,368	0	3,430	0	15,673,424	23.6
1919	0	380,651	0	396	0	0	0	1,247	0	14,560,770	25.6
1920	850	451,117	332	577	0	453	0	1,472	0	12,465,156	25.3
1921	830	478,733	50	1,849	0	414	0	3,073	0	14,003,832	28.9
1922	535	346,137	32	1,819	0	2,130	0	3,358	1,075	11,119,580	23.5
1923	313	276,770	80	0	0	527	0	1,511	0	11,409,385	23.0
1924	115	325,582	0	0	0	232	0	698	0	9,317,213	18.4
1925	178	186,819	0	0	0	0	0	1,093	0	9,018,111	18.9
1926	43	122,758	313	133	0	334	220	28,000	0	9,461,843	20.5
1927	1,015	183,697	0	299	0	0	0	389	0	9,327,439	21.4
1928	61	120,525	197	230	307	951	0	1,878	0	8,616,673	19.4
1929	1,085	246,969	0	0	0	956	0	6,114	0	8,439,483	19.8
1930	994	214,419	0	980	0	0	0	3,946	134	9,101,690	20.6
1931	1,776	229,851	470	1,988	0	784	0	3,164	623	9,361,498	20.4
1932	449	334,923	521	22,639	0	3,143	0	7,999	1,637	11,254,427	23.4
1901-32 (inc.)	171,785	11,931,125	24,814	98,602	3,534	69,376	6,782	363,270	122,839	516,397,073	22.9
Through 1932	5,171,785	36,931,125	1,024,814	7,098,602	3,534	1,069,376	1,006,782	5,363,270	622,839	1,944,833,772	41.6

See page 12 for footnotes.

1776

1 Data from Mineral Resources of the United States, except as otherwise noted. Mine production is used where available rather than mint returns.

2 Two thirds of total gold production of Alabama through 1900. See Dunlop, J. P., Gold, Silver, Copper, Lead, and Zinc in the Eastern States in 1914: U.S. Geol. Survey, Min. Res. of the U.S., 1914, part I, pp. 139-163.

3 Placer-Gold Production of Alaska, 1880-1923, inclusive, by years, is given by Brooks, A. H., Alaska's Mineral Resources and Production, 1923: U.S. Geol. Survey Bull. 773, 1925, p. 9.

4 See table by Tenney, J. B., and Wilson, E. D., in Arizona Gold Placers and Placering: Bull. Univ. of Arizona, Arizona Bureau of Mines, vol. 4, no. 6, Aug. 15, 1933, p. 15.

5 California placer-gold production from 1848 to 1900, by decades, and 1901 to 1926, by years, is given by Hill, J. M., Historical Summary of Gold, Silver, Copper, Lead, and Zinc Produced in California, 1848 to 1926: Econ. Paper 3, Bureau of Mines, 1929, pp. 10-11.

6 Production of all Eastern States included under Georgia in 1904.

7 Colorado placer-gold production, 1858-67, 1867-1923, by years, is given by Henderson, C. W., Mining in Colorado: U.S. Geol. Survey Prof. Paper 138, 1926, p. 69.

8 About two thirds of total production of gold in Georgia, 1830-1900. See Dunlop, J. P., Gold, Silver, Copper, Lead, and Zinc in the Eastern States in 1914: U.S. Geol. Survey Min. Res. of the U.S., 1914, pt. I, pp. 139-163.

9 Total estimated gold production of Idaho, 1860-70, plus one half total gold production, 1871-1900. See Ross, C. P., A Graphic History of Metal Mining in Idaho: U.S. Geol. Survey Bull. 821, 1931, pp. 1-10, for data on total gold production.

10 Montana's placer production, 1862-80, was estimated at $150,000,000. See Report of Director of Mint on Production of Precious Metals in the United States, 1884, p. 286. To this has been added about one fifth of Montana's total gold production, 1881-1900. See Report of Director of Mint, 1900. For Montana placer-gold production by years, 1896-1914, see Heikes, V. C., Gold, Silver, Copper, Lead, and Zinc - Montana: U.S. Geol. Survey Min. Res. of the United States, 1914, part I, p. 772.

11 See Smith, A. M., and Vanderburg, W. O., Placer Mining in Nevada: Univ. of Nevada Bull., vol. 26, no. 8, Dec. 15, 1932, p. 10.

12 The production of placer gold in New Mexico, 1828 "to date" was estimated at $13,000,000 to $15,000,000 by Lindgren, W., Graton, L. C., and Gordon, C. H., The Ore Deposits of New Mexico: U.S. Geol. Survey Proof. Paper 68, 1910, p. 75.

13 Approximately two thirds of total gold production of North Carolina, 1799-1900. See Dunlop, J. P., work cited.

14 Approximately two thirds of total gold production of Oregon, 1848-1900, as recorded at U.S. mints and assay offices. See Diller, J. S., Mineral Resources of Southwestern Oregon: U.S. Geol. Survey Bull. 546, 1914, pp. 22-23. The estimate is probably much too low.

15 Approximately one fourth of South Carolina's total gold production, 1829-1900. See Dunlop, J. P., work cited.

16 See O'Harra, C. C., Early Placer Gold Mining in the Black Hills: Black Hills Engineer, vol. 19, no. 4, 1931, p. 361.

17 The early production of the placers in Bingham Canyon, Utah, is estimated at $1,000,000. See Butler, B. S., Loughlin, G. F., Heikes, B. C., and others, Ore Deposits of Utah: U.S. Geol. Survey Prof. Paper 111, 1920, p. 131. The same publication, p. 133, gives Utah placer gold production, 1904-17 by years.

18 Approximately one third of total Virginia gold production, 1828-1900. See Dunlop, J. P., work cited.

19 Approximately one fourth of total gold production of Washington, 1860-1900.

20 Approximately one half of Wyoming's total gold production, 1867-1900. See Henderson, C. W., Gold, Silver, and Copper in Wyoming in 1914: U.S. Geol. Survey Mineral Resources of the United States, 1914, pt. I, pp. 248-249.

21 Total United States gold production, including Alaska but excluding the Philippine Islands, 1801-1900, was 115,-275,707 ounces or approximately $2,382,960,000.

22 Total gold production of United States, including Alaska but excluding the Philippine Islands, 1901-32, was 109,-117,309 ounces or approximately $2,255,660,000.

On the other hand, a sudden increase in the load of the stream from above, or a raising of base level, may bury the gold-bearing gravels under many feet of barren sediments; or, as in California, volcanic activity may bury gravels under beds of lava or ash. Hundreds of miles of such buried channels have been traced and mined in the Sierra Nevada. The gravels of these deeply buried placers are typically "tight" owing to the pressure of overburden or are cemented by the deposition of lime or silica in the voids. Further erosion may expose such deposits, and sometimes exceedingly rich placers are formed in the new channels. In California the present drainage lines are distinct from the old ones but likewise run generally westward and bear sufficient similarity so that the names of some of the present streams have been applied to the old Tertiary rivers.

Where rivers empty into main valleys or lakes they form alluvial fans or deltas, and extensive placers may be built up; sometimes deposits are continuous from one river mouth to the next. Mertie[20] speaks of such placers as coalescing placers. They also may either be elevated or buried by subsequent events.

Where gold-bearing material is discharged into the ocean, beach deposits sometimes are formed. None has been of great importance in the Western States. The few rich ones were small, and since their exhaustion the lower-grade deposits have been the basis chiefly for intermittent large-scale projects all of which have failed. The famous beach deposits at Nome, Alaska, are typical examples of this class of placer. These were phenominally rich. Both at Nome and on the California-Oregon-Washington coast elevated beach deposits occur several miles inland.

A few other types of placers merit description, although they are of less relative importance. Very rich gold lodes may weather and erode so as to leave a valuable deposit of gold in place in a mass of disintegrated gangue material. This type is known as a residual placer. The broken mass of rock may gravitate slowly down a steep hillside, assisted by frost and by trickling rain waters, thus bringing about a low and imperfect concentration of gold on the bedrock, and forming an eluvial placer. A deposit of this type is the placer at Round Mountain, Nev.,[21] where in 1908 operators with two hydraulic monitors were reported to be washing out $20,000 per month. Current operations at Round Mountain are described in a subsequent paper.[22] The gravel of such deposits is typically angular, very poorly assorted, and generally loose.

Glaciers may play a part in the formation of placers, as at Breckenridge, Colo., where the valley terrane or river wash lying downstream from the terminal moraines has been profitably dredged for many years. In the Weaverville district, Trinity County, Calif., a glacial till that covers large areas is known to be gold-bearing although not sufficiently so to be of economic interest. It is believed, however, to be one important source of the gold of the present and earlier channels of the Trinity River and its tributaries.[23]

Certain of the "dry" placers of the Southwest are so different from typical river deposits as to deserve a separate classification. The gold, instead of occurring in well-defined channels of old or present streams, is distributed in poorly assorted, angular, or subangular gravels over large areas of gulches, hillsides, mesas, and ridges. Usually there is only slight concentration in present gulches, and in some deposits there is little con-

---

20 Mertie, J. B., The Occurrence of Metalliferous Deposits in the Yukon and Kuskokwim Regions: U.S. Geol. Survey Bull. 739, 1923, pp. 160-162.

21 Ransome, F. L., Round Mountain, Nev.: U.S. Geol. Survey Bull. 380, 1909, pp. 44-47.

22 Gardner, E. D., and Johnson, C. H., Placer Mining in the Western United States: Part II. - Hydraulicking, Treatment of Placer Concentrates, and Marketing of Gold: Inf. Circ. 6787, Bureau of Mines, 1934.

23 MacDonald, D. F., The Weaverville-Trinity Center Gold Gravels, Trinity County, Calif.: U.S. Geol. Survey Bull. 430, 1910, pp. 48-58.

centration on bedrock. Most of the gold is relatively coarse. These placers are believed to be mainly the result of erosion by floods. Torrential rains produce sudden heavy flows of water that pile up noticeably and often prominent alluvial fans along the mountains at the mouth of each canyon. The water commonly overflows its channel and spreads over the entire fan. The position of the channel itself often is greatly changed during a single flood. The flood waters diminish rapidly, at last depositing a thin coating of silt over the whole area. It is obvious that the channels, if changed frequently, will not be particularly rich, that the swift water will carry most of the fine gold away, and that the opportunity for sorting the gravels and concentrating the gold on bedrock will be too short to be effective. The ground, in fact, may scarcely be wet before the water has stopped flowing. An occasional surface concentration has been noted, which is difficult to explain unless it is a result of wind action.

An unfortunate characteristic of the dry placers from the miner's viewpoint is the common lime-cemented condition of the gravels. This "caliche" often is as hard to mine as and much more water-resistant than any of the "tight" or cemented Tertiary gravels of California.

The thickness and general character of the gravels at individual mines operated in 1932 are given in the chapters on mining methods in the three papers of this series.

The location of the placer-mining districts in the Western States is shown in figure 1.

CHARACTERISTICS OF PLACER GOLD

Placer gold occurs as particles ranging in size from minute grains to nuggets weighing 100 or 200 pounds. Pieces worth more than 5 or 10 cents are spoken of as nuggets; smaller ones are "colors." A scale of sizes, quoted from C. F. Hoffman by Lindgren,[24] is as follows:

Coarse gold, plus 10-mesh.
Medium gold, minus 10-plus 20-mesh.
Fine gold, minus 20-plus 40-mesh.
Powder (flour) gold, minus 40-mesh.

Here, the medium gold averaged 2,200 colors per ounce, or, if pure and valued at $35, about 2/3 of a color to a cent; the fine gold, 12,000 colors per ounce or 3 colors to a cent; and the powder, 40,000 colors per ounce or 10 colors to a cent. Most beach gold and some river gold, such as that of the Snake and Green Rivers, is much finer, ranging from 200 to 1,000 colors to a cent.

Colors and even nuggets almost always are flattened to some extent. Some placer gold occurs as thin flakes, which makes recovery more difficult as the flakes are not separated readily by water action from the more compact rounded grains of heavy minerals such as magnetite or garnet.

Placer gold occurs universally as an alloy with silver. Ordinarily it ranges in fineness from 700 to 950 parts of pure gold to 1,000 parts of the natural alloy, the remainder being chiefly silver. However, lower and higher degrees of fineness are common. Lindgren[25] cites the Folsom dredging field, Sacramento County, Calif., where the gold ranges from 974 to 978 fine. Yale[26] states that some gold from a drift mine near Vallecito, Calif., was 993

24 Lindgren, Waldemar, Tertiary Gravels of the Sierra Nevada of California: U.S. Geol. Survey Prof. Paper 73, 1911, p. 67.

25 Lindgren, Waldemar, work cited, p. 68.

26 Yale, C. G., Gold, Silver, Copper, Lead, and Zinc in California: U.S. Geol. Survey, Min. Res. of the U.S., 1910, pt. I, p. 365.

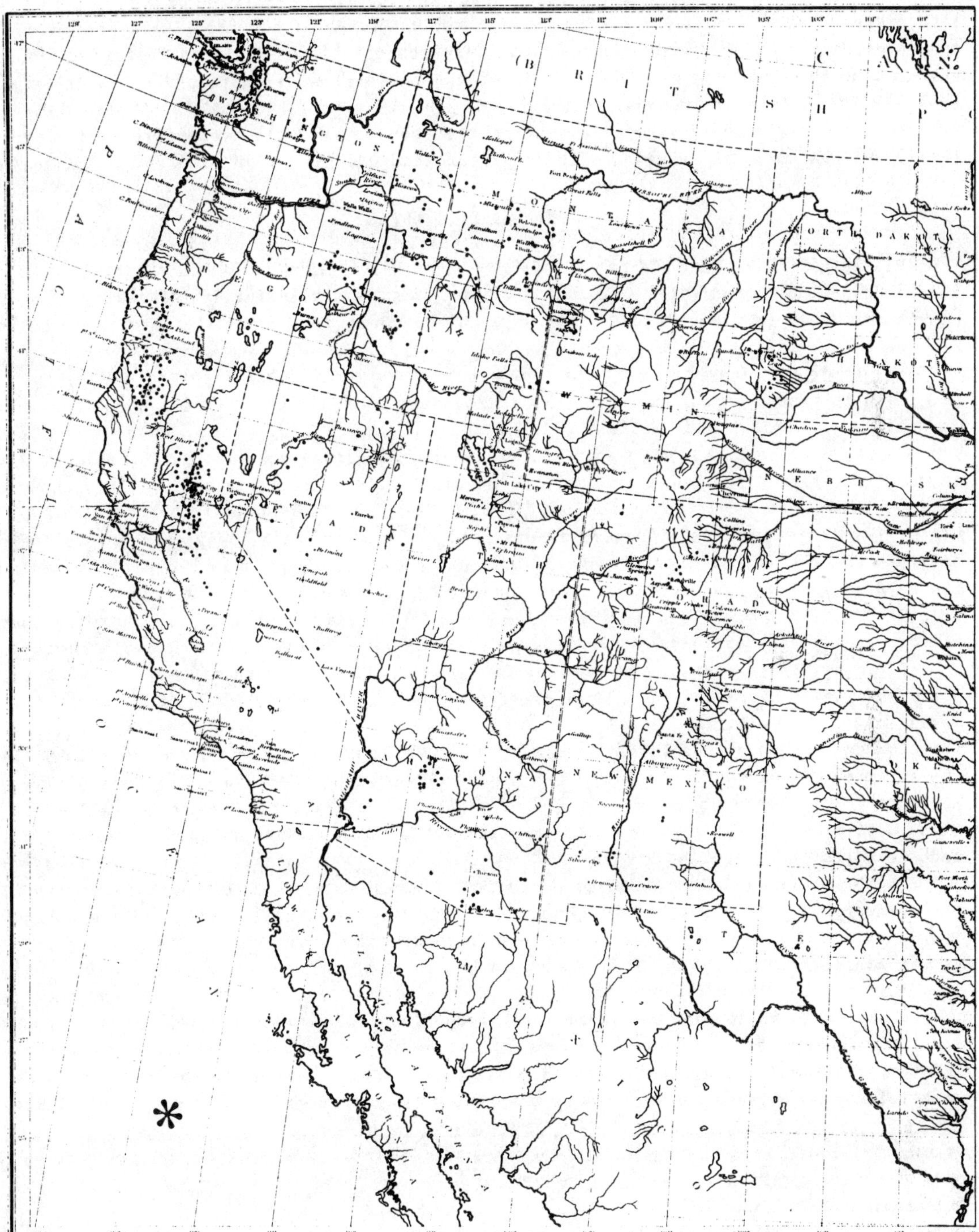

FIGURE 1. - Location of Placer-Mining Districts in the Western States.

fine, or $20.52 per ounce (at $20.67), and that the gold from this property never fell below 955 fine. In a single small district the fineness of the gold is fairly uniform for any one channel. Some miners consider the fineness a distinguishing feature of a channel in districts where several channels are being explored or mined. This rule, however, is subject to exceptions, because varied sources of gold may contribute to a placer deposit and because the gold appears to lose part of its silver content and hence increases in fineness as it travels farther from its source. According to several authorities this is due to dissolving of the silver by surface waters, an action that would have relatively more effect on fine particles than on large nuggets. Fine or flour gold usually is of relatively high purity.

Associated Minerals

Placer gold invariably is accompanied by other heavy minerals, which comprise the black, white, or yellow concentrates found in the sluice box when cleaning up. Table 3 lists these minerals roughly in the order of their commonness. Some of the characteristics noted under Remarks apply chiefly to the minerals as they are found in sluice-box concentrates.

Magnetite.- Magnetite is by far the commonest mineral constituent of the heavy sands and often is a serious nuisance to the miner because it tends to pack in the riffles. It is almost always present and frequently is as much as 1 or 2 percent of the entire weight of gravel. In some beach and river-bar placers where the mining practice is to "skim" the rich streak, magnetite often amounts to several percent of the material washed. Although a valuable iron ore under certain conditions, magnetite is worthless as a byproduct of placer mines. The impression held by some miners that it contains gold is erroneous. No instance of physical or chemical natural combination of gold and magnetite is known to the authors. The gold content of placer concentrates consisting of magnetite and other heavy minerals is practically all in the form of loose particles of gold. Some of the gold, however, may be attached to quartz particles or other gangue material, particularly if near its source.

Titanium Minerals.- Ilmenite and rutile are the chief ore minerals of titanium but do not have commercial value as gold placer-mine products. It was formerly believed[27] that the black sands of the Pacific coast, which in places contain considerable of these minerals, constituted economically valuable reserves of iron, platinum, and titanium. Recently titanium has been produced commercially from the beach sands of California.[28] Probably no gold is recovered in this operation. Rutile is worth only a few cents per pound in the form of 94-percent concentrate; the demand is supplied amply by some 200 tons produced annually from lode deposits in Norway and Virginia.

Garnet.- Garnet is an abundant accessory mineral in many rocks, particularly if they are metamorphic. It has a higher specific gravity than the usual rock-forming minerals and is present in the concentrates from sluice boxes in many localities. The gem varieties of garnet are the commonest and least valuable of the semiprecious stones; they are used much in cheap jewelry, but the production of gem garnet in the United States, which was never large, has declined to such an extent that production figures are no longer kept. Garnet is one of the most useful abrasive materials, and a considerable production is so utilized. However, this production is from lode deposits exclusively, and no instance is known of the commercial production of either gem or abrasive garnet from gold placers in this country.[29]

---

27 Day, D. T., and Richards, R. H., Useful Minerals in the Black Sands of the Pacific Slope: U.S. Geol. Survey, Min. Res. of the U.S., 1905, pt. I, 1906, pp. 1175-1258.

28 Youngman, E. P., Deposits of Titanium-Bearing Ores: Inf. Circ. 6386, Bureau of Mines, 1930, pp. 3-5.

29 Myers, W. M., and Anderson, C. O., Garnet: Its Mining, Milling, and Utilization: Bull. 256, Bureau of Mines, 1925, 54 pp.; also, Aitkens, I., Garnets (Gem Stones): Inf. Circ. 6518, Bureau of Mines, 1931, 11 pp.

TABLE 3.- Chemical composition and physical characteristics of the chief heavy minerals found in gold placer gravels[1]

Mineral	Chemical formula	Color	Specific gravity	Hardness	Remarks
Gold	Au (+Ag)	Gold-yellow	15.6-19.3	2.5	Very malleable and ductile.
Magnetite	Fe_3O_4	Iron-black	5.2	5.5-6.5	Shiny grains; strongly magnetic.
Ilmenite	(Mg,Fe) TiO_3	do.	4.5-5	5-6	Only faintly magnetic; affects compass needle slightly.
Garnet	$R''_3R'''_2(SiO_4)_3$[2]	Red, brown, various	3.8	6.5-7.5	Vitreous luster; generally in rounded crystals (dodecahedrons).
Zircon	$ZrSiO_4$	Brown, pale yellow, or colorless	4.7	7.5	Adamantine luster.
Hematite	Fe_2O_3	Dark steel-gray to iron-black	4.9-5.3	5.5-6.5	Particles generally smooth, rounded, often red-coated.
Chromite	$FeCr_2O_4$	Iron-black to brown-black	4.1-4.9	5.5	Sometimes feebly magnetic; brown streak.
Olivine	$(Mg,Fe)_2\ SiO_4$	Olive-green	3.3-3.4	6.5-7	Good cleavage; vitreous luster; transparent to translucent.
Epidote	$HCa_2\ (Al,\ Fe)_3\ Si_3O_{13}$	Pistachio-green	3.2-3.5	6-7	Distinct cleavage.
Pyrite	FeS_2	Pale brass-yellow	4.9-5.1	6-6.5	Usually cubic grains; brittle; metallic luster.
Monazite	(Ce, La, Di) PO_4 + ThO_2	Yellow	4.9-5.3	5-5.5	Resinous or greasy luster; usually in rounded grains.
Limonite	$2Fe_2O_3.3H\ O$	Dark brown	3.6-4.0	5-5.5	Yellow-brown streak.
Rutile	TiO_2	Red-brown to red	4.2	6-6.5	Distinct cleavage; metallic-adamantine luster.
Platinum	Pt (usually also Fe,Ir,Os)	Whitish steel	16.5-18	4-4.5	Malleable; sometimes scales and grains.
Iridium	Ir (Also Pt, etc.)	Silver-white, yellow tarnish	22.6-22.8	6-7	Usually in angular grains; no cleavage.
Iridosmine	Ir, Os	Tin-white to light steel-gray	19.3-21.1	6-7	Usually in flat grains; slightly malleable to brittle; good cleavage.
Cinnabar	HgS	Red	8-8.2	2-2.5	Scarlet streak.
Wolframite	(Fe, Mn) WO_4	Black, dark gray	7.2-7.5	5-5.5	Submetallic luster; good cleavage in one plane.
Sheelite	$CaWO_4$	White, pale yellow, brown, or gray	5.9-6.1	4.5-5	Adamantine, greasy luster; translucent.
Cassiterite	SnO_2	Brown or black	6.8-7.1	6-7	Brittle; rounded grains.
Corundum	)				
Sapphire	) Al_2O_3	Blue, red, yellow, brown	3.9-4.1	9	Adamantine to vitreous luster.
Ruby	)				
Diamond	C	White, colorless, pale	3.5	10	Adamantine or greasy.
Mercury	Hg	Tin-white	13.6		Small opaque fluid; silvery globules.
Amalgam	Hg, Ag, Au	Silver-white	13-14		Brittle to malleable; rubs silvery coat on copper.
Galena	PbS	Lead-gray	7.4-7.6	2.5-2.7	Metallic luster; lead-gray streak; perfect cubic cleavage; friable.
Silver	Ag	Silver-white	10.1-11.1	2.5-3	Malleable and ductile; tarnishes black.
Copper	Cu	Copper-red	8.8-8.9	2.5-3	Ductile; malleable.
Bismuth	Bi	Silver-white	9.8	2.5	Sectile; brittle; metallic luster.
Cerussite	$PbCO_3$	Colorless or white	6.5	3-3.5	Adamantine luster.
Columbite-tantalite	(Fe,Mn)(Nb,Ta) $_2O_6$	Iron-black to gray or brown-black	5.3-7.3	6	Brilliant to submetallic luster; often iridescent; brittle; good cleavage.
Quartz	SiO_2	Colorless	2.6	7	No cleavage; vitreous to greasy luster.
Feldspars	Silicates of K,Na,Ca,Al,etc.	Colorless, white, pale yellow, or pink	2.5-2.7	6-6.5	Good cleavage; vitreous luster.

1 Dana's Textbook of Mineralogy, 4th ed., by W. E. Ford, 1932, was used for most of the mineralogical data.

2 R'' = Ca, Mg, Mn, or Fe; R''' = Fe, Al, or Cr.

Zircon.- Zircon[30] has the highest specific gravity of all gem stones; in addition, certain characteristics of hardness, high refractive index, and color are slowly making it more popular as a gem. Its frequent occurrence in sluice-box concentrates is therefore of interest. It is said to be abundant in the gold-bearing gravels of Henderson County, N.C., and crystals of it are common in most auriferous sands; in some workings it comprises several percent of the concentrates. The crystals are seldom of gem quality, and the usual material is worthless as a zirconium ore because of the low price of the latter - about 10 cents per pound for a semirefined product. Even as a gem uncut zircon brings only a few dollars per carat and seldom over $15 when cut.

Hematite.- Hematite often is found in placer gravels, particularly in the dry placers of New Mexico and Arizona. Sometimes hematite cobbles and small pebbles are said to indicate rich ground, but this idea is likely to be misleading.

Chromite.- Chromite occurs very commonly in black sands, in some deposits greatly exceeding the magnetite in quantity, bu has no commercial value as found in gold placers.

Olivine and Epidote.- Olivine and epidote are common dark, heavy, rock-forming minerals and consequently are often present in the concentrates They are of no value.

Pyrite.- The occurrence of pyrite sometimes has mineralogical interest because it is believed to have formed in place in gold placers. It is often found in the Tertiary gravels in California and sometimes is thought to have resulted from the reducing action of organic matter on iron sulphates in the meteoric waters.[31] Pyrite is also found as detrital grains or masses derived from lode deposits or from the country rock.

Monazite.- Monazite, a phosphate of the rare-earth metals, valuable because of its varying small content of thorium, is characteristic of the Boise Basin placer sands of Idaho but is found in many other districts in Idaho, Colorado, Montana, and Oregon. It is notably lacking in the California deposits, with rare exceptions. A serious attempt was made about 1909 to utilize the monazite contained in the tremendous quantities of hydraulic tailings lying in the Boise Basin fields.[32] A 1,000-ton plant was built, comprising tables, drier, screens, and several magnetic concentrators. A 95-percent monazite sand could be made, containing about 5 percent thoria. None was ever marketed, probably because the cost would not permit competition with the much richer Brazilian and Indian beach deposits.[33]

Limonite.- Limonite is an iron-ore mineral found occasionally in placer concentrates and has no economic importance in this connection.

Platinum and Osmiridium.- Platinum and osmiridium are found in many districts and are almost the only consistently valuable byproducts of American gold-placer mines. They are typical of the beach-placer sands of California and Oregon, where they occur in proportions as high as a tenth or more of the gold.[34] Practically all the Sierra Nevada placers contain platinum-group metals, as do the placers of the Klamath Mountains, particularly those on the Trinity River in the Hay Fork and Junction City districts.[35] None of these deposits, however, have contained enough platinum to be worked for that metal alone. Indeed, the dredge operations are the only ones to produce platinum consistently, even as a byproduct of gold mining, the total annual output seldom being over 300 or 400 ounces. In a recent year one

---

30 Youngman, E. P., Zircon (the Gem): Inf. Circ. 6465, Bureau of Mines, 1931, 20 pp.

31 Lindgren, Waldemar, Tertiary Gravels of the Sierra Nevada: U.S. Geol. Survey Prof. Paper 73, 1911, p. 76.

32 Sterrett, D. B., Monazite and Zircon: U.S. Geol. Survey, Min. Res. of the U.S., 1909, pt. II, 1911, pp. 898-903.

33 Santmyers, R. N., Monazite, Thorium, and Cerium: Inf. Circ. 6321, Bureau of Mines, 1930, 43 pp.

34 Hornor, R. R., Notes on the Black Sand Deposits of Southern Oregon and Northern California: Tech. Paper 196, Bureau of Mines, 1918, 42 pp.

35 Day, D. T., Platinum: U.S. Geol. Survey, 19th Ann. Rept., pt. VI, vol. 1, 1897-1898, pp. 265-271.

dredging company, which recovered about 50,000 ounces of gold from 10,000,000 cubic yards of gravel, sold about 110 ounces of crude platinum metals; these contained 70 ounces of platinum and 6 or 8 ounces each of iridium and osmium. Very few hydraulic or other placer operators find it worth while to separate the platinum from the gold because of the small quantity involved and the difficulty of finding a purchaser. Although platinum sold for as much as $154 per ounce in 1920[36] its average price in 1932 was only $36.46 per troy ounce of refined metal, and its average price in January 1933 was $26.48. In January 1934 the price was $38.00 per ounce.

The prices of iridium and osmium fluctuate widely, iridium having sold during recent years for $50 to $300 per ounce and osmium for $25 to $115.

Cinnabar - Cinnabar is found in sluice-box cleanups in a few localities. It is unmistakable because of its brilliant red color. Edman[37] states that minute quantities of cinnabar, in grains and crystals, are found in the platinum-bearing sands of Plumas County, Calif. It was seen recently by the junior author of this paper in pan concentrates from the gravels of Copper Basin Wash near Skull Valley, Ariz., in the form of round pellets the size of bird shot. Cinnabar is known to occur in lode deposits in that locality.

Tungsten Minerals.- Wolframite and scheelite, the most important tungsten-ore minerals, occasionally are found in placer deposits. In Alaska they have been mined from placers in which they were associated with gold. In the States, particularly in the Atolia-Randsburg district of California and in the Boulder County tungsten district, Colo., considerable tonnages of tungsten minerals have been recovered from eluvial or residual placer deposits.[38] At Atolia scheelite was mined and milled on a commercial scale until about 1930 when operations ceased due to the low price of tungsten.[39] In Colorado the mineral was ferberite, the high-iron member of the wolframite series. Huebnerite, the high-manganese end member of the series, has been mined by dry-washing during periods of high prices from shallow alluvial or residual deposits at Round Mountain, Nev.[40]

Cassiterite.- Cassiterite, the principal ore of tin, has been recovered from placer deposits in Alaska and has been noted in a few placer deposits in the States; small quantities have been mined in the Appalachian gold-tin district. None of the occurrences in the Western States has proved to have commercial importance.

Corundum.- Sapphire and ruby are varieties of the mineral corundum; they are translucent or transparent, of a fine blue or red color, and otherwise of gem quality. When the mineral is an opaque light blue, brown, or gray it is known as corundum, and if granular and mixed with impurities such as magnetite it is called emery. Neither corundum nor emery can be produced by placer mining to compete with richer and better-situated deposits or with the increasing production of artificial aluminum oxide abrasives; but sapphires and rubies, especially if of gem quality, are valuable.

It is not known that sapphires or rubies have ever been of much value in this country as byproducts of gold-placer operations. However, in several districts in Montana, notably along the Missouri River near Helena, sapphires were discovered in the course of gold mining, and several attempts were made in 1891 and later years to wash the bars for the gems. These ventures were financial failures, partly because the investment cost was high and partly be-

---

36 Tyler, P. M., and Santmyers, R. N. Platinum Inf. Circ. 6389, Bureau of Mines, 1931, 69 pp.

37 Edman, J. A., The Platinum Metals of Plumas County, Calif.: Min. Sci. Press, vol. 77, Oct. 22, 1898, p. 401.

38 Hess, F. L., Tungsten Minerals and Deposits: U.S. Geol. Survey Bull. 652, 1917, pp. 44-45.

39 Vanderburg, W. O. Methods and Costs of Concentrating Tungsten Ores at Atolia, San Bernardino County, Calif. Inf. Circ. 6532, Bureau of Mines, 1931, 12 pp.

40 Ferguson, H G., The Round Mountain District, Nevada: U.S. Geol. Survey Bull. 725, Contributions to Econ. Geol. 1921, pt. 1, 1922, p. 389.

cause the sapphires were not of good color. In 1899-1900 about 25,000 carats of gems, including a few small rubies suitable for cutting, were selected from about 400,000 carats hydraulicked from the gravels of Rock Creek, Granite County, Mont. Beginning about 1905 these mines were again worked, the chief product being sapphires of a quality suitable for watch jewels and other bearings. Occasional gem stones also were found, and a small quantity of gold was saved as a byproduct. The average selling price of the bearing sapphire was about $1 per ounce. Since 1917 the mines have been operated intermittently on a reduced scale.

From about 1905 to 1915 a bucket dredge was operated intermittently in another Montana gold-sapphire district, on Cottonwood Creek, Deerlodge County. The gulch gravels here were 10 to 14 feet thick and were composed of porphyry cobbles and subangular boulders with 3 or 4 feet of black muck overburden.[41] Both gold and sapphires were concentrated near bedrock. The gold was fine and was said to pay operating expenses. Most of the sapphires were of a quality suitable only for mechanical purposes, although some gem stones were found.

Montana has continued to produce sapphires from the Judith Basin district lode mines. One Montana lode-mining company in 1924, 1925, and 1926 produced an average of about 50,000 carats of sapphires suitable for cutting into gems less than 1 carat in weight. This output was valued at about 50 cents per carat, and the same company's much greater production of industrial stones was valued at a small fraction of a cent per carat.[42] (One carat equals 200 mg.) Large gem sapphires, on the contrary, have sold for as much as $1,500 per carat.

Diamonds.- Diamonds have been reported from gold-placer gravels in California and other districts. Lindgren[43] states that the principal localities in California are Cherokee Flat, Butte County, where 56 specimens, ranging up to 1 1/2 carats in size, were reported, and Placerville, Eldorado County. Tyler[44] gives the placer mines of the Volcano and Fiddletown districts, Amador County, as the chief producers. The gold-bearing gravels of south central Indiana occasionally produce diamonds; one weighing 1 1/2 carats has been described.[45]

It is believed that diamonds originate in the same rocks as platinum -- namely, the serpentines resulting from alteration of peridotites. The diamonds found in gold placers in this country have not been of commercial importance, being small, yellowish, and far from abundant. Doubtless many pass over the riffles because of their relatively low specific gravity (less than 40 percent greater than quartz), and probably many more pass unrecognized into the rejects from the treatment of concentrates. In Arkansas, where diamonds are mined by hydraulic methods, the principal recovery is made on grease-covered vibrating tables.

Quicksilver.- The natural origin of quicksilver in gold placers seldom can be established definitely, because very little strictly virgin ground is now being mined in the United States. Dana's Textbook of Mineralogy states, however, that it does exist in some alluvial deposits. In view of the frequent occurrence in lode deposits of native quicksilver associated commonly with one of the mercury sulphides and the finding of cinnabar in placer deposits, it is not unreasonable to assume that some of the metallic mercury recovered is of natural origin. It is certain, however, that as a rule such quicksilver has been introduced in earlier mining activities. Dredges and hydraulic mines frequently recover considerable mercury in their riffles. In the Skull Valley district of Yavapai County, Ariz., the authors were informed that one plant excavating and treating the shallow gravels of certain

---

41 Sterrett, D. B., Precious Stones: U.S. Geol. Survey Mineral Resources of the United States, 1907, part II, 1908, pp. 821-822.

42 Aitkens, I., Rubies and Sapphires: Inf. Circ. 6471, Bureau of Mines, 1931, 11 pp.

43 Lindgren, Waldemar, Tertiary Gravels of the Sierra Nevada of California: U.S. Geol. Survey Prof. Paper 73, 1911, p. 75.

44 Tyler, P. M., Abrasive and Industrial Diamonds: Inf. Circ. 6562, Bureau of Mines, 1932, p. 14.

45 Schaller, W. T., Gems and Precious Stones: U.S. Geol. Survey Mineral Resources of the United States, 1916, part II, 1917, pp. 892-893.

tributaries of Copper Basin Wash had recovered enough mercury to increase considerably its original stock and to furnish a neighboring plant and a number of "snipers" with several pounds of mercury. This district has been the scene of intermittent 1- or 2-man operations for several decades, yet the occurrence of cinnabar in the gravels makes the natural origin of metallic mercury at least plausible.

Amalgam.- Amalgam is recognized as a mineral only when it is a crystalline combination of mercury and silver having fairly definite characteristics and is thus described in mineralogical textbooks. Clarke,[46] however, gives two analyses of native gold combined with mercury. One from Colombia contained approximately 84 percent of gold, 8 percent of silver, and 7 percent of mercury; the other specimen, described as amalgam from Mariposa County, Calif., contained 39.02 percent of gold and 60.98 percent of mercury (approximating the formula $AuHg_3$, and had a specific gravity of 15.47. Most of the amalgam recovered in sluice boxes, however, undoubtedly is of artificial origin.

Galena.- Galena is an uncommon constituent of placer concentrates, presumably because it is both soft and friable and because it oxidizes readily. It was observed in sluice boxes by the authors in the Cedar Creek district of Montana; in some specimens it still adhered to part of the original matrix of gangue minerals.

Other Minerals.- Although all placer gold is alloyed with silver, no record has been found of the occurrence of native silver, as a separate metal, in placer deposits in the United States, and silver is included in the list only for comparison with gold. It is known, however, that many silver nuggets were found in the Nizina district of Alaska; one nugget, with attached quartz, weighed over 7 pounds.[47]

Copper nuggets have been found in placer gravels in many districts. Like the reported occurrence of bismuth in the gravel of French Creek, Summit County, Colo., this fact is of mineralogical interest only, at least so far as any district in the Western States is concerned. The placers of Chititu Creek in the Nizina district, Alaska,[48] contain enough copper to be a nuisance (several hundred pounds at each clean-up). It is not known that any of this was ever marketed.

In California Gulch, above Leadville, Colo., where placer mining preceded the discovery of valuable lead deposits by several years, cerussite (lead carbonate) was known to the miners simply as a variety of rock that was so heavy as to obstruct sluicing.

The mineral columbite-tantalite occurs in some pegmatite veins. The Black Hills region of South Dakota furnishes fine specimens of this mineral, and it is reported, together with cassiterite and scheelite, in the placer gravels of Bear Creek, in the northern Black Hills.

Quartz and feldspar, which constitute the bulk of placer sands, are included at the end of table 3 for comparative purposes. It will be observed that their specific gravities are one fourth lower than those of olivine and epidote, a third lower than that of common garnet, half those of magnetite and hematite, and only a sixth or seventh that of gold.

Other minerals besides those listed in table 3 are found in sluice-box concentrates, some because of their heavy weight and some merely because of imperfect concentration. Many artificial objects likewise find their way into the concentrates, the typical ones being bird shot and nails -- quite often the hobnails of the "oldtimers."

---

46 Clarke, F. W., The Data of Geochemistry: U.S. Geol. Survey Bull. 770, 5th ed.. 1924, p. 658.

47 Moffitt, F. G., and Capps, S. R., Geology and Mineral Resources of the Nizina District, Alaska: U.S. Geol. Survey Bull. 448, 1911, p. 105.

48 Moffitt, F. G., and Capps, S. R., work cited, pp. 105-106.

LOCATION OF PLACER CLAIMS ON PUBLIC LANDS

Placer claims containing alluvial deposits of gold or other metals can be located and patented on the public domain, national forests, stock-raising homesteads, and unpatented parts of congressional grants to railroads. Public land temporarily withdrawn from settlement, location, sale, or entry and reserved for water-power sites, irrigation, classification, or other public purposes shall at all times be open to exploration for metalliferous minerals and purchase under the mining laws. However, power or reservoir sites withdrawn by congressional action or Executive order are not subject to mineral location.

Although placer claims can be located and mineral rights obtained on stock-raising homesteads, written permission must be received from the homesteader to enter upon the land, or a bond of $1,000 must be posted to indemnify the agricultural entryman for any damage that may be done to the crops or tangible improvements. Surface rights are limited to the land actually needed for mining purposes.

Mining claims cannot be filed upon patented land except where the minerals have been reserved to the United States, on military or naval reservations, or in national parks or monuments. Land below high tide, lake beds (except Searles Lake, Calif.), or the beds of navigable rivers are not subject to mineral location.

Public land, in the public-land States, valuable for minerals cannot be patented except under the provisions of the mining law, and valid mineral locations take precedence over other forms of land entry.

According to the Federal law, mining locations, both lode and placer, may be made by citizens or those who have declared their intention to become citizens, by an association of qualified persons, or by a domestic corporation. Locations can be made without regard to age, sex, or residence of the locator. No limit is placed by the Federal statutes on the number of locations that may be made in the United States by an individual or a company. A locator may include as colocators other persons who may or may not have seen the ground; also, a person may make valid locations for other parties.

The Federal statutes require that a location notice must contain the names of the locator or locators, the date of location, and a description of the claim by reference to some natural object or permanent monument that will identify the claim. A discovery of valuable mineral within the limits of the claim must be made for the location to be valid.

State laws in most placer-mining States define how the location notice must be posted, the size of the discovery shaft or other discovery excavation that must be sunk if required, and the claim boundaries must be marked. These laws require that the location must be filed with the proper county official.

A single placer claim located by an individual locator is limited to a maximum of 20 acres. The law, however, permits the location of association claims. That is, an association of 2 individuals can file on a 40 acre claim; 3 individuals on 60 acres, etc., up to a group of 8 persons who can locate 160 acres as a single claim.

One discovery of mineral is required to support a placer location, whether it is 20 acres of an individual or 160 or fewer acres of an association of persons. Placer claims should conform as nearly as possible to legal subdivisions of the public land surveys, except where the rectangular subdivisions would necessitate placing the lines upon previously located claims. The smallest legal subdivision is a square tract containing 10 acres and measuring 660 feet each way. Although it has been held that placer claims may be located to conform to their environment where the topography is such that it is impractical to lay out rectangular claims, as in gulchs with precipitous walls, the Land Office regulations require that the entries be as compact as possible and will not permit entries for patent which cut the public domain in long, narrow strips or grossly irregular tracts.

Both lode and placer locations can be amended at any time and the boundaries changed, provided such changes do not interfere with the rights of others.

A placer claim cannot be located over an older lode location if the lode claim is located legally on a deposit of valuable mineral in place. The owner of a valid lode location also owns any placer deposits the claim may contain.

A lode claim can be located legally on known lodes in place containing valuable mineral on an unpatented placer claim held by the lode locator or others in the same manner as if the placer location did not exist. A lode deposit cannot be held under a placer location, but once a placer claim is patented the owner owns and may mine all lodes not known to exist at the time the patent was issued.

Although valid mineral entries have precedence over agricultural or other entries, a placer claim must be shown to be more valuable for minerals than for agriculture in contested cases.

To hold the possessive title to a mining claim not less than $100 worth of work must be performed or improvements made upon or for the benefit of each claim each year, regardless of its size. Where a number of contiguous claims are held in common, the aggregate expenditures for the group may be made on one claim. Locations connecting only at the corners are held to be noncontiguous. The period within which the annual work must be done commences at noon of July 1 succeeding the date of location. Failure to do the annual assessment work will subject a claim to relocation unless work is resumed before such relocation. It has been held that a claim is not subject to relocation if work is being performed on the ground at the end of the required period. In other words, if work is begun on a claim located, say, in September 1928 by noon of July 1, 1930 and diligently carried on thereafter until complete, it is not subject to relocation. Additional work would be required for the period commencing July 1, 1930. Annual expenditure is not required subsequent to making entry at the Land Office for patent. Congress, by the act approved June 6, 1932, relieved claim owners of the necessity of doing the annual assessment work on unpatented mining claims for the period ended July 1, 1932. Congress by the act approved May 18, 1933 also relieved claim owners (except those required to pay the Federal income tax) of doing the annual work for the year ending July 1, 1933.

The annual assessment work may be omitted on a claim for 1 or more years, and the location will still be valid if work is resumed on the ground, provided no interfering interests are affected or no other location has been made on the ground.

Where one of several locators fails to contribute his share of the required expenditures made for the benefit of a claim the co-owners at the expiration of the year may give notice personally or in writing or by advertising in the newspaper published nearest the claim at least once a week for 90 days; if upon the expiration of 90 days after the personal notice or upon the expiration of 180 days after the first newspaper notice the delinquent co-owner shall have failed to contribute his proportion of such expenditures or improvements, his interest in the claim passes by law to his co-owners who have made the required expenditures.

The Secretary of the Interior has been authorized by Congress (act of June 30, 1929) to lease unallotted lands on Indian reservations for mining purposes in Arizona, California, Idaho, Montana, New Mexico, Oregon, Washington, and Wyoming. After declaration by the Secretary that the lands are subject to lease, claims may be located as on the public domain; a duplicate of the location notice must be filed within 60 days with the superintendent in charge of the reservation. The locator has 1 year's preference right to apply for a lease through the reservation superintendent to the Secretary of the Interior. Leases are for 20 years, with provision for 10-year renewals.

Most of the mining States make provision for leasing minerals found on State lands. After discovery, application for a prospecting or mining lease should be made to the authority having charge of State lands. Regulations regarding the granting of prospecting or mining leases vary in different States.

PROSPECTING OUTFITS AND PROVISIONS.

The outfit to be taken on a prospecting trip depends upon the mode of transportation, the work contemplated, and the funds available. Enough equipment should be taken, but unnecessary articles make extra work. When a more or less permanent camp is established, added equipment for personal comfort and efficiency can be obtained. Usually a cabin is built for a permanent camp.

Camp Outfit

Most prospectors these days use automobiles and can carry complete camp equipment.

A tent should be carried if an extended trip is to be made. Few prospectors, however, put up a tent for one night unless the weather is bad. Bedding consists of 3 or 4 blankets rolled up in canvas; the number of blankets needed depends upon the climate. A folding cot is advisable; however, in a permanent camp a bunk is usually built. A full-size single-bitted ax should be carried; the one-handed "Boy-Scout" ax is practically useless. A saw and a hammer with 2 or 3 pounds of assorted nails is needed in fixing up a camp and for other purposes. All prospectors and campers carry a good-quality, stout pocket knife. A miner's acetylene lamp provides a good light; a 5-pound can of carbide will last a camp nearly all summer. A flashlight is very convenient, but an ample supply of batteries must not be overlooked.

A 50-foot length of 1/2- or 5/8-inch manila rope is useful for lashing the load to the car or truck, as a tent rope, for getting down into old shafts or cuts, and in making a "Spanish winch" for pulling the car out of a mudhole or for moving heavy equipment. To make this winch, one end of rope is attached to the object to be moved and the other to a tree or other anchor some distance ahead. A 5- or 6-foot length of stout sapling is held vertically or stood in a hole midway between the ends. A loop formed in the middle of the rope is placed around the end of a long pole or pipe which is used as a sweep to wind the rope around the sapling, thereby shortening the rope and pulling the object ahead.

A canvas sheet is handy to protect the load from rain and dust. A canvas war bag is needed for clothing and other articles. A tight, stout, wooden box with a lid saves food by keeping insects and rodents away from it and is a great convenience in making and breaking camp.

A stove is necessary if a camp is to be maintained in cold weather. A combination heating and cooking stove is best, preferably one with an oven in which bread can be baked.

Camp furniture usually is improvised on the spot, the kind and amount depending upon the length of time the camp is to be used or the camper's idea of personal comfort.

A shotgun and a few boxes of shells can be used to provide a change in diet where rabbits or game birds abound; the cost of the ammunition, however, may exceed the money value of the food thus obtained. Many prospectors living in the hills keep rifles to supply the larder with venison. A few fishhooks and a line, or even more elaborate tackle, take little room in the car and provide relaxation and food where streams contain fish. However, too much sporting equipment interferes with the purpose of a prospecting trip, and guns probably would prove a nuisance as they cannot be left unguarded in a camp without the likelihood of their being stolen.

A 2-quart canteen with a shoulder strap usually is needed for carrying drinking water or water for panning. A 2-gallon canteen and a 5- or 10-gallon water keg are necessary in some districts.

Clothing

The most important item of clothing is a pair of stout, thick-soled shoes of good quality, preferably hobnailed. If an extensive trip is planned a second pair may be needed. Other clothing can be patched, but when a prospector's shoes go to pieces his trip is ended. A pair of rubber boots will prove a comfort if much placering is done.

Woolen socks to wear under the heavy shoes help to prevent blisters; several pairs may be worn out in a season.

Other clothes are chosen for the climate and service. A leather jacket is very serviceable and comfortable in cool weather, or a sheepskin coat may be needed when it gets colder. Many prospectors in mountainous regions wear flannel shirts and woolen underwear. Overalls are a common garb.

A complete change of clothing should be taken on all except the shortest trips to permit changing into dry clothing after being caught out in the rain or working in water all day.

Tools

A pick, a long-handled, round-pointed shovel, a gold pan, and a prospector's pick are indispensable. If claims are to be staked, a compass will be needed for running out the lines. A hand magnifying glass is a great help in identifying minerals. If lode deposits are to be sought a mortar and pestle, a horn spoon, or a small pan will be needed for testing rock for free gold or other heavy minerals. A blowpipe outfit and determinative tables are of service to those who can use them. Bags for taking out samples usually are needed. Double paper bags with rubber rings cut from old automobile tubes for closing the bags permit large numbers of samples to be collected with little expense for bags.

A single-jack hammer with 2 or 3 moils will come in handy for taking samples and for loosening rock encountered in making cuts.

Some prospectors carry 1 or 2 sets of hand steel and a few pounds of powder. A few rounds may be drilled and blasted before the steel has to be resharpened. If any extensive rockwork is to be done a forge and a set of blacksmith tools are necessary; these usually are brought in later.

If any placering is contemplated a rocker and enough lumber to build one or two 12-inch sluice boxes may be carried.

Cooking Equipment

For a 1- or 2-man party, a frying pan, a coffee pot, a large and a small stew pan or pot, and a Dutch oven are needed. A knife, fork, spoon, cup, and plate are required for each man. A few extra plates come in handy. A good butcher knife, a water pail, a can opener, and a few tea towels complete the set. Other dishes can be taken according to personal preferences.

Provisions

The variety of food taken on prospecting trips depends upon the method of transportation and the prospector's pocketbook. If the supplies are to be packed on animals, bulky foods such as potatoes and canned articles are omitted. If there is need for economy in making purchases, the list will consist mostly of dried staples and vegetables if available locally.

Bacon, flour, beans, oatmeal, dried or canned fruit, coffee, syrup for hot cakes, and sugar and canned milk for the coffee are the stand-bys in prospectors' camps. As funds get low more beans and less bacon are eaten, and canned fruit is omitted. Canned tomatoes are commonly used; they are cheap and supply needed food elements not contained in dry staples. Where available locally Irish potatoes, onions, and other vegetables are eaten. Fresh meat is not used much in camps in the summer on account of the difficulty of keeping it. In dry climates, however, a quarter of beef can be eaten by a crew of a dozen men before it spoils by using proper care. The meat is hung in the open air during the night and in the morning, while still cool, is wrapped in blankets or a bed roll for the day.

A proper balance should be made in making out a "grub" list so that needed items will not run short. Everyone has preferences for different articles of food which should be followed as much as practicable. It has been found by experience, however, that fancy groceries are the ones left over and the first articles to be used are the bacon, potatoes, and flour. Plain, wholesome fare seems to be preferred in camp, especially where hard work is done.

The old United States army ration for one man for 1 day was: 12 ounces bacon or pork, or 22 ounces fresh beef; 18 ounces soft bread or flour, or 16 ounces hard bread, or 20 ounces corn meal.

For each 30 rations the following items were issued: 5 pounds beans or peas, or 10 pounds rice; 5 pounds sugar; 1 quart vinegar; 1/4 pound hard candy; 1 pound soap; 1 1/2 pounds salt; and 1 1/4 ounces pepper.

The following standard ration list for fireguards is used as a guide by the Forest Service. The articles listed are enough for one man for 30 days.

Article		Quantity
Bacon, salt	pounds	2
Bacon, smoked	do.	10
Baking powder	pound	1
Baking soda	do.	1/2
Beans, pink	pounds	5
Beef, fresh (purchased locally)		...
Butter, best	pounds	2
Candles		6
Cheese, full cream	pounds	1 1/2
Chili powder	small bottle	1
Coffee, best	pounds	4
Corn	cans	6
Dried fruit, variety	pounds	9
Flour	do.	24
Green chili	cans	3
Jam	jar	1
Lard	pounds	4
Matches	large boxes	2
Milk	tall cans	10
Oatmeal	pounds	6
Onions, dry	do.	5
Peas, best grade	cans	6
Pepper, black	pound	1/4
Potatoes, Irish	pounds	15
Raisins, seedless	do.	2
Rice	do.	3
Salt, table	pound	1

Article	Quantity
Sauerkraut..........cans	3
Soap, hand..........cake	1
Soap, laundry..........do.	1
Spinach..........cans	3
Sugar..........pounds	12
Syrup..........gallon	1
Tomatoes..........cans	6
Towels, tea..........	3

The following weekly allowance of food for one person to give a balanced diet is condensed from suggestions made by Doctor Smith:[49] Three 1-pound cans evaporated milk; 2 pounds potatoes; 4 pounds onions, cabbage, beets, or other vegetables; 3 pounds citrus fruits, or 6 pounds fresh apples, or equivalent dried prunes, apricots, etc.; 3 pounds dried beans; 6 to 8 pounds cereals, whole-wheat flour or bread, rolled oats, shredded wheat, etc.; 2 1/2 pounds dried meat, bacon, ham, or cheese (fresh meat or eggs may be substituted if available); 3 pounds sugar; 1 pound coffee; 1/4 pound salt; 1/2 pound butter; and baking powder.

In many districts of the Southwest water must be carried. The quantity required depends upon the time of year and the amount of work done by the miner. Men working where temperatures range from 100° to 110° drink 2 gallons or more of water per day; under such conditions, a 10-gallon tank would last one man 3 days, allowing for cooking but not for the radiator of the car. In cooler weather a 10-gallon tank should last one man a week or 10 days.

First-Aid Supplies

As prospectors are likely to be away from medical aid, some medical and first-aid supplies should be taken along. These should consist of a laxative (castor oil or salts), iodine or mercurochrome to disinfect cuts or bruises, and a first-aid kit. A snake-bite kit may also prove invaluable.

SAMPLING AND ESTIMATION OF GOLD PLACERS

Failure to sample and estimate properly the available yardage of placer deposits has resulted in a tremendous waste of money and effort. A large proportion of all placer operations has failed because the gold in the gravel was insufficient to repay the cost of even the most efficient mining, not to mention the return of money invested or interest thereon.

Many methods of sampling are available, including the simple panning of gravel from natural exposures, drifting, test-pitting or trenching, shaft-sinking, and churn-drilling. Actual mining on a small scale often is done as a method of sampling prior to investing considerable money in development or equipment. Several examples will be noted later under methods of mining.

The technique of panning and its use for estimating the gold content of gravel are discussed under the head of Panning and Rocking, as these operations are properly considered small-scale mining methods. In the following section the more elaborate methods of sampling are discussed, and costs are given where available.

49 Smith, Margaret Cammack, Food Suggestions for Prospectors, Arizona Gold Placers and Placering: Univ. of Arizona Bull., vol. 3, no. 1, Jan. 1. 1932, pp. 96-98.

Weight of Placer Gravel

Placer gravels vary greatly both in weight per cubic yard in place and percentage increase in volume on being loosened. Yet in making estimates of yardage and value it often is necessary to use some factor to convert volume in place into loose volume or into tonnage. In common placer terminology "heavy" gravel indicates coarse rather than weighty material. The weight is greater in tight or cemented than in loose ground, and it increases with the proportion of large boulders and heavy rock material such as diorite, greenstone, or hornblende schist. The amount of moisture present likewise affects the weight.

Three contiguous samples taken by the authors from the same bed of tight, fine clayey gravel overlying a pay streak in the Greaterville district, Arizona, indicated weights of 3,450, 3,540, and 3,000 pounds to the cubic yard, respectively, or an average of 3,300 pounds. A sample of clean gravel with 30 to 40 percent cobbles over 2 1/2 inches in diameter, from another gulch in the same district, weighed 3,600 pounds to the cubic yard in place. The samples contained 5 to 8 percent of moisture. The expansion of the first three samples was 50, 54, and 33 percent, respectively, an average of 46 percent. The last sample expanded 17 percent.

The gold-bearing river gravel at the pit of the Grant Rock-Service Co., Fresno, Calif., weighs 2,850 pounds per cubic yard. Some engineers in calculating tonnage allow 3,000 pounds to the cubic yard, bank measure. Handbooks give weights per cubic yard ranging from 2,600 to 3,650 pounds. An average weight probably is between 3,000 and 3,300 pounds to the cubic yard.

Sampling Natural Exposures

Whenever a gold pan is used the result not only proves or disproves the presence of gold but also usually shows accurately the amount of gold contained in the sample chosen. Panning along a creek bed thus is the most elementary method of sampling a placer deposit. However, the samples commonly are taken so as to render the quantitative results worthless.

A gravel deposit of much size seldom can be sampled directly from its natural exposures; but a few creek banks, steep-sided gulches, or old excavations such as hydraulic pits may be available, in which case certain precautions should be taken to get true samples. First, the vertical extent of gravel to include in a given sample should be determined. If hydraulicking is to be done, the whole depth of gravel ordinarily is included in one sample, except when it is planned to pipe off the barren overburden to waste, in which event it is desirable to know the depth of barren material and samples may be taken of each distinguishable stratum. If drifting is planned, only the lower, economically minable gravel need be sampled. After the location and extent of a sample cut have been decided, care must be taken to have equal quantities of gravel from all points along its length. The best way to do this is to cut a channel or groove of uniform shape and size from top to bottom of the sample distance. Enough such samples must be taken to prove the continuity of the "pay streak."

Mechanically the procedure of sampling a gravel face resembles closely that pursued in lode mines. A pan and a pick are the requisite tools. The bank should be trimmed well and cleaned along the sample line to eliminate effects of surface weathering. The pan may be used to catch the loosened material, or a canvas may be spread on the ground. If conditions favor it, a measurable channel or groove should be cut so that the volume taken can be measured. Otherwise, the only alternatives are to use a factor for pans per cubic yard or to measure the loose gravel in a box which has been calibrated in terms of bank measure.

Drifting

Drifting is a common method of prospecting a deep placer deposit when conditions are favorable. Methods and costs of drifting are discussed briefly under Drift Mining in a subsequent paper.[50] The cost of driving a small drift in placer gravel ordinarily ranges from $2 to $6 per foot. Difficult ground conditions or excessive water may increase the cost to $10 or $15 per foot.

For sampling purposes the gravel usually is taken to the surface and concentrated in sluice boxes. If an old drift is being sampled various methods may be followed. If the ground will permit, the most satisfactory method is to slab 1 or 2 feet from the side of the drift and wash the gravel thus broken. If not, vertical channels may be cut on one or both sides of the drift at intervals of about 5 feet. If the latter is done, the volume of the sample cut may be measured, which is preferable under most conditions, or a factor may be used for reducing loose measures of gravel to solid measure, which facilitates taking the sample but introduces some uncertainty and leads to carelessness in sampling. If values are to be expressed in cents per ton it is still necessary to decide what conversion factor to use in making estimates of tonnage.

Test-Pitting

Test pitting and trenching are applicable only to gravels so shallow that a man can throw out the dirt by hand. The best procedure is to mark out the area of the pit on the surface, making it rectangular, as small as convenient, and preferably in dimensions of even feet such as 2 feet wide and 3 or 4 feet long. Then it should be excavated to bedrock with smooth, vertical walls. Sometimes a cleared space is prepared and the dirt thrown on the bare ground, but in view of the greater difficulty and possible small error involved in rehandling the dirt it is better to shovel it onto a canvas or board platform or into a receptacle. If a large boulder projects into the pit no correction of the theoretical volume of gravel should be made, regardless of whether or not the boulder is removed or allowed to remain in place, as obviously it cannot be ignored in mining operations and is the equivalent of so much barren gravel. The gravel taken from the pit may be thrown into one or more piles, depending on whether or not information is desired regarding one or more individual strata. Sometimes alternate third, fourth, or tenth shovelfuls are used for the sample to reduce the volume to be panned or otherwise concentrated. This should be avoided whenever practical because of the possible error introduced.

The cost of sinking test pits or running trenches in earth and gravel has been noted often enough for fair generalizations to be established. Gardner[51] states that opencast work in placer ground costs from $0.40 to $1.00 per cubic yard, depending on the nature of the ground and on wages. Wages for such work then ranged from $3 to $4 per 8 hours. Furthermore, a man should be able to pick and shovel about 8 cubic yards of fairly loose gravel in 8 hours. In test pits the worker's efficiency would be lowered somewhat by the cramped quarters and by the care necessary to square out the corners and trim the sides to vertical planes.

Near Skull Valley, Ariz., an area of shallow gulch placer ground was being sampled by test pits in 1932. The gravel was moderately fine and loose, angular to subangular wash, 2 to 6 feet deep. Mexican laborers at $3 per day were able to dig an average of five test pits

---

50 Gardner, E. D., and Johnson, C. H., Placer Mining in the Western United States: Part III. - Dredging and Other Forms of Mechanical Handling of Gravel, and Drift Mining: Inf. Circ. 6788, Bureau of Mines, 1934, 81 pp.

51 Gardner, E. D., Cost of Mine Openings: Eng. and Min. Jour., vol. 100, Nov. 13, 1915, pp. 791-794.

per 8-hour man-shift. The pits averaged 2 feet wide, 3 or 4 feet long, and 3 1/2 feet deep. The cost of digging the pits was therefore about $0.65 per cubic yard.

The gravel was shoveled first onto the ground beside the pit. Usually, all gravel was hauled by truck to the washing plant, an average distance of about three fourths mile. If a pit produced more than about 1 yard of gravel the sample was reduced in size by quartering. The sampling was done near a well on Copper Basin Wash. The gravel was shoveled from the truck directly into the hopper of a small screening, washing, and concentrating device mounted on a trailer. A 1-inch centrifugal pump on the trailer took about 15 gallons of water per minute from a large tank on the hillside and forced it through the sprays of a double trommel screen, in which the gravel was washed; all material over about one eighth inch in size was rejected. The fines were concentrated in a patented 12-inch centrifugal bowl. The plant was said to have a capacity of 2 to 3 yards per hour. It was driven by a 1 1/2-hp. gasoline engine. Water was pumped from the well to the storage tank by a 3 1/2-hp. gasoline engine.

About 12 men were employed - 1 at the washer, 2 on each of two trucks, and 7 or 8 digging pits. Wages were $3 per day. About 25 samples were taken and washed per day or 470 during a period of about 3 weeks. The total cost was $1,000, itemized as follows:[52]

Labor	$705.70
Trucks	96.00
Lumber	12.85
Sacks, canvas, etc.	24.50
Supplies and repairs for washing plant	25.75
Gasoline, oil, and grease	53.35
Rental of water tank and pipe	15.00
Miscellaneous supplies and expense	66.85
Total	1,000.00

The spacing of test pits depends on the nature of the deposit. If the pay gravel occurs in narrow channels the best plan is to space the pits in lines across the channels, as is done with churn-drill holes when sampling dredge ground. The holes must be close enough to yield an average value which represents the average value of the channel at that point. In practice the spacing ranges from a few to 50 or 75 feet, depending on the uniformity of results. The transverse lines of pits theoretically should be placed close enough to show either fairly uniform values from line to line or a reasonable upward or downward trend of gold content along the channel. Unfortunately, this is seldom possible, and the usual practice of spacing the lines from 100 to several hundred feet apart is a compromise in the engineer's mind between the cost of sampling and the need for accurate results.

Shaft-Sinking

Shaft sinking is the usual method of testing placer ground. Prospect and sampling shafts, unless intended for later use in drifting or mining and unless exceptionally deep (75 feet or more), are sunk as small as practicable. The usual section is rectangular and 3 by 4 to 4 by 6 feet in size. Round timber 4 to 6 inches in diameter which is available in most districts, is commonly used to crib shafts in loose gravel. In gravel tight enough to stand safely without lagging the only timber necessary is stulls set to hold the ladder.

---

52 Information supplied by W. R. Shanklin, Oatman, Ariz.

A hand windlass is the usual means of hoisting, using a light steel, 2-cubic-foot bucket and a 3/4- or 1-inch-diameter manila rope, as ordinary wire rope is unsuitable for a windlass; 75 to 100 feet is the maximum depth at which such equipment can be used most efficiently. For greater depths a power hoist of some kind should be installed. Large boulders can be lifted with the ordinary hand windlass if it is provided with long cranks; as much as 800 pounds can be raised by two men. Such feats, however, are considered dangerous because of the general absence of safe brakes or catches on windlasses, and the possibility of killing or injuring the operator if he loses control of the crank handle.

On Bear Gulch, north of Bearmouth, Granite County, Mont., a 4- by 6-foot shaft was sunk to a depth of 32 feet in 1932 at a cost of $3 per foot. The shaft was largely in fine, mucky, loose, angular gravel. It was cribbed solid with 6-inch round timber. A crew of three men averaged 6 feet per day. Wages were $3.50. Little or no water was encountered, and no blasting was required as the ground could be loosened by pick.

On Sauerkraut Creek in the Lincoln district, northwest of Helena, Mont., one man working alone sank a 4- by 5-foot shaft through dry, loose, hillside wash and moderately compact bench gravel at the rate of 3 feet per day to a depth of 30 feet, cribbing the shaft solidly with small round timber.

A 100-acre tract of possible dredging ground in the Pioneer district, Powell County, Mont., was sampled in 1931 or 1932 by sinking shafts to bedrock at 400-foot intervals, or one shaft to 3.6 acres.[53] The gravel was moderately fine, lacked large boulders, and was compact enough to stand without support in 4-foot-diameter shafts, except through a 4- or 5-foot surface accumulation of hydraulic tailings. The depth to bedrock was 10 to 65 feet and averaged 35 feet.

A crew of four men contracted to sink 1,000 feet of shaft at the rate of $1 per foot. They worked very hard and on long shifts, with the result that their earnings averaged $3.50 per day. Two of the men worked individually at starting shafts and putting in the collars and enough board lagging to hold back the loose surface wash. When the shaft became too deep to throw material out by hand the digger would leave it and start on the next. Then the other pair of men installed a hand windlass and sank the shaft to bedrock. The shaft at the surface was cribbed about 4 by 4 feet square; but a short distance down the cross section was converted to circular, with a 4 1/2-foot diameter. Under favorable conditions two men would sink 10 feet per day.

The sampling was done later by an engineer and three assistants. A portable windlass was used, which could be moved in a truck from shaft to shaft. The engineer cut all samples, taking a channel 4 inches wide and 2 to 5 inches deep straight down one side of each shaft. The loosened dirt was caught on a canvas placed in the bottom of the shaft. Distinct strata were sampled separately; the bottom few inches of rather coarse material was relatively very rich. The surface tailings which were consistently barren were not sampled. It was not necessary to dig into bedrock which, although soft, was barren. Usually 3 or 4 samples were taken per shaft. The crew of four men with a truck could sample three 40-foot shafts per day.

Once or twice, where a shaft had caved in before being sampled, it was necessary to sample the dump. This was done by hand-shoveling, throwing every tenth shovelful onto a canvas on which it was rolled and quartered.

The sample was measured in a box about 18 inches square and 18 inches deep, which was calibrated against a measured excavation in solid gravel. It was found that this gravel expanded about 20 percent upon being dug. Each sample finally was washed in a rocker. One rockerman, with the engineer in charge acting as waterman, could wash and clean up about 10 samples per day, averaging about 1/10 cubic yard each.

---

53 Information given by A. V. Corry, under whose direction the sampling was done.

This ground was sampled very cheaply. The cost for the shaft sinking was only $10 per acre. The contract of $1 per foot could not be duplicated in normal times. The contractors, however, had almost no expenses except food, as all the timber needed was picked up locally for little or nothing; and their tools consisted of shovels, hammers, and picks. No blasting was done. The sampling cost is not known but from the data supplied could be estimated on a basis of ordinary labor costs as perhaps $0.40 per foot of shaft or $4 per acre.

Many shaft-sinking costs are available in the literature of placer mining. According to Janin,[54] they range from considerably less than $2 to as high as $25 per foot. He cites the instance of a Colorado dredging field, where with labor at $1.50 to $1.75 per day, more than 100 shafts 5 to 30 feet deep were sunk at a cost, excluding sampling, of 24 cents a foot. Representative shaft-sinking costs are given and commented on by Bigelow in the section on churn-drill sampling.

An excellent description of a placer examination is contained in a recent article by D. L. Sawyer.[55] The property was a typical desert dry placer in the Weaver-Rich Hill district, Arizona. The gold of the district ranged from very coarse to fine but averaged about flaxseed size. The gravel was loose and contained so many large boulders that sampling was difficult. It was decided therefore to sink shafts and wash all or most of the gravel excavated. The shafts averaged 3 1/2 by 5 1/2 feet in section and 12 feet in depth. Most of them had to be cribbed, which was done with framed, 2- by 12-inch plank. The shafts were sunk a few inches into bedrock which was cleaned carefully with a whiskbroom and scoop. The cost of shaft sinking was $3 to $6 per foot with hired labor but on contract was $2 per foot for the first 10 feet and $3 for the second 10. The contractors were said to have earned $5 per day.

The dirt from each 5 feet of shaft ordinarily constituted a sample, except where the interval was shortened to separate different strata of gravel. Boulders more than 6 inches in diameter were thrown aside, and the remainder of the gravel was shoveled into 1-cubic-foot wooden boxes and hauled by truck to the washing plant. Careful measurements of the shaft at each sample interval gave the dimensions needed to calculate the yardage of the sample. It was found that in ground lacking boulders over 6 inches in size 1 cubic yard filled 50 boxes, indicating an expansion of 85 percent, which may have been caused partly by failure to fill the corners of the boxes.

The washing plant, said to have a capacity of 6 to 10 yards per day, consisted of a shoveling platform, a hopper and set of screens, a piped water supply from a near-by mine, and a line of four sluice boxes. The boxes were 8 feet long, 10 inches wide, and 6 inches high, made of 1-inch, clear, surfaced lumber, and set in steps so that the material from one dropped into the next. This arrangement obviated the necessity for tight joints and permitted ready adjustments of the grade of each box. The upper box was fitted with a 4-foot riffle consisting of a 1-inch board with 1-inch holes bored at 3-inch intervals. The second and third boxes had transverse riffles 1 1/2 inches high and 8 inches apart. Burlap covered with expanded metal lath was used in the fourth box. The grades of the boxes were adjusted so that the riffles "boiled" properly and the black sand, which averaged about 20 pounds per ton, did not pack; in sluicing one sample the slopes were 1 1/8 inches per foot for the first, second, and third boxes and 1/2 inch for the fourth. The grades were altered to give the best recoveries with different types of gravel. No quicksilver was used. In a run made with a 25-cubic-foot sample, previously tested and found barren, then loaded with about 30 grams of placer gold, 60 percent of the gold was recovered in the first box, 36 percent in the second and third, and less than 0.2 percent in the fourth; about 4 percent was lost.

---

54 Janin, Charles, Gold Dredging in the United States: Bull. 127, Bureau of Mines, 1918, p. 30.

55 Sawyer, D. L., Sampling a Gold Placer: Eng. and Min. Jour., vol. 133, July 1932, pp. 381-383.

The gravel was shoveled by hand onto the upper screen where it was washed by a flow of 50 gallons per minute coming through a perforated 2-inch pipe. Plus 1-inch material was raked off at this point and plus 1/2-inch material, from the lower screen, the balance falling onto the apron of the sluice, whence the flow of wash water carried it over the riffles. A 1/4-inch screen was used for treating clayey gravel in place of the 1/2-inch screen to insure complete disintegration.

Concentrates from each clean-up were panned and dried; the black sand was separated from the fine gold by a magnet, followed by blowing on a smooth surface or by amalgamation. After the gold was weighed on a balance to the nearest 0.01 mg, that from each sample was bottled and marked with a label showing the number of the shaft, depth in feet represented, volume in cubic yards, weight in milligrams, and value in cents.

At Gold Gulch near Bowie, Ariz., about 200 prospect pits were sunk in 1932 and 1933 to determine the yardage, nature, and value of an area of typical dry-placer ground. The deposit ranged from 3 or 4 to about 30 feet in depth and consisted of a few inches to 2 or 3 feet of clayey soil underlain by 1 to 6 feet of caliche (lime-cemented gravel), which in turn was underlain by tight, angular to subangular gravel containing a large percentage of rock and many boulders 6 inches to 2 or 3 feet in diameter. The bedrock was granite. No water was encountered in the test pits.

A gasoline-driven, 1 1/4-cubic-yard shovel already on the property was used to scoop out a trench at the site of each pit; the trench was 4 or 5 feet wide and about 5 feet deep at the center. From this point a 2 1/2- by 4 1/2-foot shaft was sunk to bedrock. Very little blasting was done, most of the gravel being picked, shoveled into small buckets, and hoisted by a hand line. No timbering was necessary. The shafts were placed 100 feet apart in lines spaced at 300-foot intervals along the gulch. No systematic sampling was done while the pits were being dug. When shaft-sinking was finished samples were taken by cutting vertical channels in one end and one side of each pit.

The work was started on company account, but later the contract system was substituted and progress was more rapid. The power-shovel work was done on a basis of 26 1/2 cents per vertical foot of excavation, the contractor furnishing the fuel for the shovel. Sinking the pits was paid for at the rate of $0.75 per foot to a point 9 feet below the bottom of the shovel cut (that is, the approximate limit of throwing out material by hand-shoveling) and $1.50 per foot from that point to bedrock. Some of the contractors earned $6 per 8-hour shift.

Exploration of the type of placer deposit that is worked by drift-mining often involves sinking one or more shafts. On Quartz Creek, near Rivulet, Mont., in 1932, two men sank a 5- by 7-foot prospect shaft 12 feet in 6 days, cribbing it solidly with 6-inch split cedar. The gravel was chiefly fine, well-washed material, with few boulders too large to handle without blasting. The first shaft was made large enough to use as a development shaft if pay gravel was found. At a depth of 12 feet, where sinking was halted by a considerable flow of water, an ingenious pumping plant was built. Power was provided by a homemade, overshot water wheel and transmitted through an endless wire-rope drive to the pump, which was set up at the collar of the shaft. The water wheel was of wood and steel, 5 feet in diameter and 3 feet wide, and was said to be capable of developing about 15 hp. The pump consisted of rods working in a 2-inch pipe, connected at the bottom of the shaft to a 4-inch pump cylinder with check valve and strainer. The pump rod was given a 10-inch stroke by an arm connected to one crank pin of an old 4-cylinder automobile engine from which the cylinders had been removed. As a separate pump unit could be driven from each crank pin, provision was made for three more pumps should they be required.

The shaft was completed to bedrock at a depth of 18 feet, requiring a total of 15 days, or thirty 8-hour man-shifts. After some drifting on bedrock three more prospect shafts were sunk in a search for the bedrock channel. These were 5 by 5 feet in the clear, cribbed solidly like the first, and 14, 18, and 18 feet deep, respectively. In this work the same crew averaged 2 feet per day.

Before completing no. 1 shaft it was necessary to install a second pump column; the two units then handled easily a flow of about 30 gallons per minute. In no. 2 shaft only one pump was needed, but in no. 3 shaft a flow of about 75 gallons per minute made it necessary to add a third pump and increase the speed to 58 strokes per minute. In sinking no. 4 shaft two pumps with a speed of 32 strokes per minute handled the water. When crosscutting was begun a third unit was added, the speed was increased to 40 strokes per minute, and finally the fourth unit was added. It was found best to keep the speed of the pump low.

A much deeper-lying deposit near Angels Camp, Calif., was prospected by both drilling and underground work.[56] The latter necessitated a shaft 167 feet deep, the sinking of which is described in a subsequent paper[57] under Drift Mining. It cost $39.50 per foot.

Drilling

The sampling of placer deposits by drilling, or, more precisely, drive-pipe sampling, is discussed in detail by Janin.[58] He states that the Keystone no. 3 traction machine is used generally in California. The usual casing is 6 inches inside diameter and 3/8 inch thick, in 5- to 7-foot sections. Drilling without casing is not good practice, and high values in a hole so drilled cannot be accepted. The bit and stem weigh 800 to 1,000 pounds. The cutting shoe usually is about 7 1/2 inches in outside diameter. Theoretically this dimension is the diameter of the cylinder of material excavated, hence it determines the yardage per foot of hole drilled. A sand pump is used to lift the loosened gravel or sludge from the hole, usually after each foot of drilling when in pay dirt. The casing is driven ahead of the tools except when boulders or cemented gravel prevents. The casing is recovered after finishing the hole.

The Empire drill has been used sparingly in this country; in foreign fields, however, it has been used extensively. It is a man-power rig, consisting of a light string of tools working inside a heavy casing (usually of 4-inch pipe) that is fitted with a toothed cutting shoe. Men or animals at the end of a long sweep turn the casing, which sinks into the ground under the weight of four men standing on a platform attached to its upper end and revolving with it. In firm ground this process is hastened by the use of a heavy weight or "jar" operated by the men on the platform. A spring pole may be used to carry the weight of a long string of tools. A sand pump is used to bring the sample to the surface. Peele's Mining Engineer's Handbook (p. 327) cites costs of $1 to $3 per foot for this work. The great advantages of the Empire drill are its light weight, low first cost, and general adaptability to remote regions and unskilled labor. These advantages are offset by the relatively smaller and less accurate samples obtainable with a 4-inch cutting shoe compared with a 6- to 8-inch shoe and by its inability to perform satisfactorily in deep or very tight gravels. With these drills 5-foot samples commonly are taken, and the sludge is measured for correction purposes in a box calibrated on a basis of 25- to 50-percent expansion of the gravel.

In an article on dredging and resoiling in Japan, Little[59] notes that the area being dredged was prospected by Empire drills, with holes at the corners of 360-foot squares. This

---

56 Steffa, Don, Gold Mining and Milling Methods and Costs at the Vallecito Western Drift Mine, Angels Camp, Calif.: Inf. Circ. 6612, Bureau of Mines, 1932, pp. 6-7.

57 Gardner, E. D., and Johnson, C. H. Placer Mining in the Western United States: Part III. - Dredging and Other Forms of Mechanical Handling of Gravel, and Drift Mining: Inf. Circ. 6788, Bureau of Mines, 1934.

58 Janin, Charles, Gold Dredging in the United States: Bull. 127, Bureau of Mines, 1918, pp. 30-49.

59 Little, H. S., Japanese Gold-Dredging Enterprise: Eng. and Min. Jour., vol. 130, Nov. 24, 1930, pp. 513-514.

is equivalent to 1 hole to 3 acres. The maximum depth of dredging ground was 33 feet. The sampling evidently was considered sufficient basis for the installation of an expensive plant, comprising a 10-cubic-foot American-built bucket dredge and an electric power plant.

Regardless of the type of drill used, the sludge sample taken from each foot or several feet of drilling usually is treated in a rocker, and the concentrates are panned to recover the gold. The colors are classified by eye according to size, usually into three groups, and a record is made of the number of each. Generally the gold is then amalgamated with a small quantity of mercury, which at the completion of the hole is dissolved in nitric acid. The gold is then weighed to give the total yield of the hole. Frequently the tailings from rocking and panning are saved and reconcentrated at the completion of the hole to detect and recover gold lost during the first washing.

Samples of placer gravel should not be assayed for gold by the usual fire methods used commercially for lode-gold ores. A pan or rocker in expert hands will recover all or more of the gold content of the gravel than can be recovered by any known device so far used successfully in placer mining.

Systematic records of the results of drilling are essential to the subsequent calculation of values and yardages. Figure 2 shows a drill log, or summary, which provided for a complete record of the information gained from a single hole.[60] It was printed on 8 1/2- by 11-inch loose-leaf sheets.

In the work on which this record form was used the panners were trained to classify the gold particles of approximately 1 mg in weight as no. 1 colors. Eight fine traces, or four coarse traces, were supposed to equal one no. 1 color. Four no. 1 colors equalled one no. 2 color, and four no. 2 colors were equal to one no. 3 color, or 16 mg. Colors heavier than no. 3 were weighed separately and calculated accordingly. In practice, the weight of the panner's no. 1 colors, as checked by the final weighing of the gold from each hole, was generally found to be between 0.75 and 1.5 mg. This classification of colors was made as a convenience in estimating the distribution of the gold through the strata. Data recorded by the driller in his drill logs comprised the times of starting and finishing, the number of different sizes of colors found at each pumping, notes on the nature of the gravels, depth measurements, and measurements of core rise and sludge volume. The last was taken by dumping the sludge into a cylindrical container of the same inside diameter as the casing and noting its height. Notations were also made of the colors obtained by repanning the tailings and by panning the settled sludge and the plug of gravel taken from the casing upon pulling it, after the hole was finished. The engineer in charge transcribed these items onto the sheet and proceeded to make his calculations. The measurement of core rise, usually about 15 percent within the casing, and of sludge volume both were used to calculate the "weighted ratio" (0.81), which indicated that an excess of material was taken from the hole; hence, the weighed quantity of gold recovered was reduced in the calculation from 31.55 to 25.55 mg. As 282 feet of hole with 4-inch casing was taken as equivalent to 1 cubic yard, the gold content of the gravel in cents per cubic yard was calculated by dividing 282 by 28.0, the depth to bedrock, multiplying by 25.55 and again by 0.0530, the value in cents per milligram of gold 795 fine (with gold at $20.67 per ounce).

A sketch of the strata penetrated was made by the engineer from the drill crew's notes and his own observation. On it were recorded the numbers of colors recovered from the gold-bearing strata, thus showing at a glance the geology and value of the gravel at that point.

Table 4 contains data on a number of placer-prospecting campaigns and was supplied by G. A. Bigelow, placer-mining engineer of San Francisco. He states:

---

60 Courtesy of V. V. Clark, mining engineer, Colorado Springs, Colo.

Placer Engineering Co.
Drill-prospecting summary

Property, Paradise Park Dredging Co.
Block no., E
Line no., C 101a
Elevation, 3,675 feet

Driller, F. R. Crawford
Drill, no. 3
Hole started May 19, 3:15 p.m.
Panner, Antonio Perez

Depth, feet	Character of ground	Number of colors: Trace	1	2	3
8.0	Yellow sandy clay, micaceous	-	-	-	-
7.5	Blue sandy clay, micaceous	2	-	-	-
1.5	Blue sandy clay, micaceous;				
	coarse granite; quartz; gravel	68	-	-	-
4.0	Blue sandy clay, micaceous;				
	fine quartz; granite; gravel	70	1	-	-
7.0	Blue shale, micaceous; quartz				
	gravel; trace bedrock	89	7	1	-
1.5	Decomposed granite bedrock	-	-	-	-
29.5					
	Repan	7	-	-	-
	Sludge	-	-	-	-
	Plug	-	-	-	-
	Total	236	8	1	

Sketch of strata

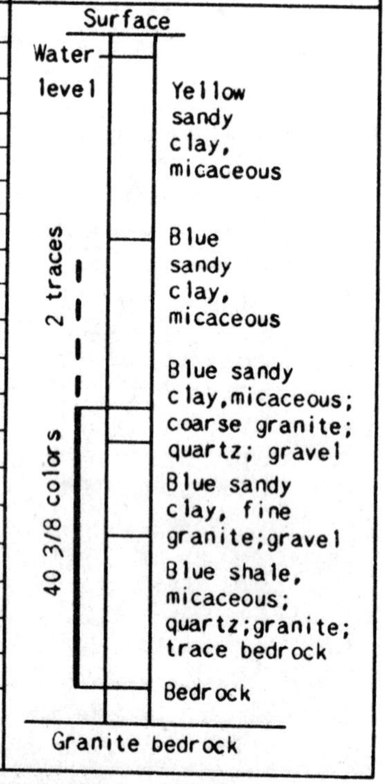

Total depth, 29.5 feet; depth to bedrock, 28.0 feet
Weight of gold, 31.55 mg
Number of no. 1 colors, 41.5
Average weight per color, 0.76 mg
Weighted weight of gold, 25.55 mg
Cents per cubic yard, 13.64
Cents per cubic yard at 85 percent recovery, 11.6

Water level, 0.5 feet
Fineness of gold, 795
Cents per milligram, 0.0530
Feet of hole to 1 cubic yard, 282
Weighted ratio, 0.81

Date, May 20, 1929

V. C. Park,
Engineer in charge.

FIGURE 2. - Drill-Log Summary Sheet.

TABLE 1.- Representative placer prospecting costs; prospecting carried to conclusion and followed by dredging except in two cases

	(1)	(2)	(3)	(4)	(5)	(6)
Style of drill	Keystone no. 3	Keystone no. 3	Keystone no. 3	Shafting	Shafting	Shafting.
Power	Gasoline-driven	Gasoline-driven	Steam	Hand; no timber	Caissons, hand-driven	Frozen; sunk by hand.
Year	1931	1931	1908, 1910	1906	1931-32	1907
Location	Folsom, Calif.	Waldo, Oreg.	Sumpter, Oreg.	Salmon, Idaho	Snelling, Calif.	Dawson, Yukon Terr.
Months	March	May-June	Aug.-Nov.; July-Nov.	June-September	November-May	June-October.
Formation	Gravel with much clay	Deep gravel with much clay	Loose gravel, 16 percent soil.	Shallow gravel	Loose gravel, 20 percent soil.	71 percent frozen muck, 19 percent frozen gravel.
Operating conditions	Favorable	Favorable	Favorable	Favorable; water pumped by hand.	Unfavorable; much water; power pumps.	Very favorable; steam thawing.
Wage rate per day:						
Foreman	$9.60	$9.60	$7.67	$7.50	$8.25	$8.00
Driller	8.25	7.25	4.50			
Panner	7.25	6.25	4.50	4.17	7.25	5.65
Helper	4.40	4.00	3.00	3.00	6.45	5.65
Assistant engineer			2.67	5.67	6.45	
Rent of team per day	8.00		5.50	6.00	6.00	4.65
Rent of truck per day	5.00	5.00	[1] .50		10.00	
Rent of drill per day	12.50	10.00	5.00			
Board, per month	35.00	37.50	30.00	20.00	37.50	50.00
Number of holes	11	16	226	51	46	64
Total depth feet	747.5	1.386	4,400.5	697	778	1,553.5
Number of crew-shifts	37.5	66	320			
Number of man-shifts				354.5	294.5	490
Number of days	22	48	160	75	194	98
Advance per shift feet	20.4	21	12.2	1.97	2.64	3.17
Advance per day do.	34	29	24.4	15.5	4.00	15.85
Average depth per hole do.	68	86.7	19.47	13.67	16.9	[2] 24.27
Costs: Labor	$1,142.55	$2,671.34		1,853.00		
Repairs	92.15	82.13				
Freight	70.00	352.65				
Travel expenses	162.67	141.14				
Drill rental	325.00	480.00	800			
Gasoline and oil	70.56	40.66				
Engineering	160.75	1,051.32		562.50		
Teams	8.00					
Rent of truck	200.00	250.00				
Total	2,231.62	5,079.24	10,169.43	2,415.50	7,666.69	5,435.47
Cost per foot	2.985	3.665	2.31	[3] 3.46	[4] 9.85	[5] 3.50

1 Wagon. 2 17 muck; 7 gravel and bedrock. 3 $2.01, sinking; $1.45, sampling. 4 $7.48, sinking; $2.37, sampling. 5 Sinking and sampling.

The prospecting was carried to a conclusion in each instance. Dredging followed the prospecting in four of the six cases.

The large difference in cost of engineering between jobs no. 1 and no. 2 was due to the much larger overhead at no. 2 on account of the nature of the deposit. The distance from headquarters at San Francisco also affected the size of the engineering force necessary on the ground. Job no. 3 was in effect two jobs. The work was interrupted for a full year.

Job no. 4 was favored by an ample force of experienced gravel miners, a low water level, and not a difficult quantity to handle. Diaphragm bilge pumps were used and in some shafts too deep for suction, a second shaft sunk 10 feet deep and adjacent to the first was used for increasing the effective depth, by means of two lifts.

Job no. 5 was a very difficult undertaking and required the use of steel telescoping caissons, especially designed for the job. Gasoline-driven pumps of the jackhead deep-well cylinder type proved very awkward but most effective.

In both jobs no. 4 and no. 5 the total contents of the shafts were washed in a long-tom device by hand and in job no. 4 a check sample was cut from the side of the shaft and washed in a rocker.

Job no. 6 is a typical case of shaft prospecting in frozen ground where the gravel deposit is unusually thin or shallow. Here the conditions for shaft work were very favorable, but the high cost of living was reflected in the unit cost. Where experienced men are available, as in this case, and equipment developed by the miners for inaccessible places is at hand for their use, a very low unit cost is obtained.

Drilling in frozen ground is also very economical, owing to the speed with which the work is accomplished and the absence of casing costs. The volume of sample is quickly and accurately obtained by water measurement after completion of the hole.

The unit cost or cost per foot for placer prospecting is usually uncertain, since it depends upon the total footage. The number of holes or cross sections as the case requires may prove to be very few, and the total cost of starting and clearing up the job falls upon a small total footage. Two cases can be cited as follows: One in Colombia, S.A., cost about $25,000 for about 1,500 feet of drilling where the equipment was left behind and never salvaged. Another in central Alaska cost about the same for less than 500 feet of drilling where the equipment was not salvaged on account of cost. In such cases and in many others that constantly arise costs can be reduced to a very low figure by a preliminary examination made by an experienced and reliable placer-mining engineer and are usually represented by the engineer's fee and expenses. The number and distribution of prospect holes needed to provide the essential information can be readily determined by the preliminary survey.

Rich, deep-lying channels, such as those in California, which are worked by drift-mining, usually are prospected and developed by drifting, because the pay streaks are too small in area and too erratic in value to be amenable to drill-sampling. However, drilling may b of great value in determining the nature of the gravel, depth to bedrock, and location and grade of the deep channel. At the Vallecito Western drift mine near Angels Camp, Calif., drilling was applied in this way. Steffa states:[61]

---

61 Steffa, Don, Gold Mining and Milling Methods and Costs at the Vallecito Western Drift Mine, Angels Camp, Calif.; Inf. Circ. 6612, Bureau of Mines, 1932, pp. 5-6.

The broad Tertiary valley east of Six Mile Creek was prospected by churn drilling to discover the location of the actual channel. A north-south line of holes was drilled across the valley, spaced as shown in table 1 and figure 1 A.[62] The holes were not drilled in regular order across the valley. The first was drilled north of the valley center in order to determine the bedrock gradient of the north rim and struck bedrock at a depth of 92 feet. This depth and the distance of the hole from the outcrop of the north rim of the channel permitted an approximate calculation of the slope of the north rim. Known elevations of the channel floor at points exposed by mining operations 2 miles west of this section indicated a depth of overburden here of about 150 feet. The second hole, drilled 281 feet to the southward, reached bedrock at a depth of 144 feet. The third hole was intended to be on the south rim and, in fact, struck bedrock at a depth of 135 feet. The fourth hole, with the information then available, was aimed for the channel trough, and the fifth and sixth were drilled to explore the width and possible pay area of channel gravels. The seventh hole was drilled north of the first hole to test for a possible split in the ancient river trough, but none was found, the hole bottoming on bedrock at 72 feet.

The drilling order, location, depth, and size of churn-drill holes are shown in the following tabulation.

Cross-section no.	Order in which drilled	Location	Lateral interval, feet	Depth, feet	Size, inches
1	7	North rim	....	72	6
2	1	do.	135	92	6
3	2	do.	281	144	12
4	5	do.	43	142	6
5	4	Trough	30	145	6
6	6	South rim	31	142	6
7	3	do.	67	135	6
....	8	Shaft site	....	187	12

The drill sludge from the first 100 feet of holes in the vicinity of the channel proper was sampled intermittently. From that depth to bedrock each 5 feet was sampled carefully. The sludge was weighed first and then passed through a rocker. The concentrate was panned and the gold weighed.

The drill rig used was an assembled outfit purchased by the company and was powered by a 10-hp. gasoline engine. The crew consisted of a runner and a helper. Six and twelve inch "Mother Hubbard" type bits were used.

All of the seven prospect holes were drilled without casing, except one where it was intended to sink a prospect shaft; the plan was later abandoned, however. Two more holes were lost, one at 82 feet, due to casing trouble, and one at 57 feet when a large quartz boulder was encountered which neither hammering nor blasting would dislodge. A tenth hole was sunk at the present shaft site, as described later. All drilling was done on company account. The total footage was 1,198 feet, drilled in 278 working days, or an average of 4.3 feet per day. The average cost was $6.11 per foot. The costs of drilling the 6-inch and 12-inch holes were practically the same.

62 Table 1 and Figure 1 A are not reproduced here.

As it happened, the estimates based on the results of this drilling and sampling were proved by later underground development to be fairly accurate. Drilling, however, should not be depended on for the exploration of such deposits, as the gold concentration is usually erratic. The recovery of gold from the drill sludge should be taken merely as an indication of the presence of ore in the stream gravels, the quantity and distribution of which must be determined more closely by drifting and finally by extraction.

The Continental Dredging Co., at Breckenridge, Colo., in 1930 and 1931 drilled about 1,500 feet of 6-inch holes, averaging 50 to 60 feet deep, in bouldery ground. The cost of this work was $2.87 per foot.

The cost of maintaining a single drill crew for sampling dredge ground at Oroville in 1932 was about $1,000 per month; the cost per foot of hole drilled was $2.50. The gravel was about 50 feet deep.

The property of the York Mining Co. near Helena, Mont., was prospected by drilling. The following data were furnished by N. C. Sheridan, the engineer in charge. The deposit was a stream placer, consisting of about 3 feet of gold-bearing gravel covered by a thick layer of black mud and soil, locally known as "beaver muck", to an average depth of more than 50 feet above bedrock. Holes were drilled 20 feet apart in transverse lines as much as 800 feet long across the bottom of the valley. Five such lines were drilled within 1 1/4 miles along the stream, with a total of 154 holes averaging 54 feet deep. A Star drill rig was used, powered by a 10-hp. gasoline engine which consumed about 1 gallon of gasoline per hour. The holes were cased with 5-inch pipe in 4- to 8-foot lengths. The drill crew consisted of a driller at $4.00 per shift, a panner at $4.50, and a helper at $3.00. An average of one hole per 8-hour shift was drilled at a cost of $0.50 per foot. The overburden was barren and so soft that the casing could be driven to the gravel stratum in about 2 hours. The holes were sunk 2 or 3 feet into the soft bedrock to insure complete recovery of all gold.

A dredging property on the Trinity River, Calif., was drilled about 1922 and again in 1928. The gravel was 10 to 45 feet deep, overlying slate or greenstone bedrock; the pay streak ranged from 100 to 800 feet wide. Large boulders were not numerous, and the gravel was relatively free from clay but was cemented in places. Along the river the gravel was saturated with water, and shaft-sinking therefore had been found impractical. The gold was fairly coarse and was not confined to bedrock but occurred anywhere from 1 to 15 feet above it. Requa states:[63]

> The property was sampled in three stages: (1) The southern end of the property was drilled by the Metals Exploration Co. prior to the time that the dredge was built in 1923. About 100 churn-drill holes were put down at that time. A few shafts were sunk but the greater part of the prospecting was done with a steam-driven Keystone portable rig, using standard prospecting casing with an inside diameter of 6 inches. The holes were spaced 125 feet apart in roughly parallel rows 750 feet apart. (2) The upper ground was prospected by the Shasta Dredging Co. in 1922, using the same rig that was used in prospecting the ground of the Metals Exploration Co. Over 60 holes were put down in rows about 900 feet apart with the holes spaced 150 feet from each other. (3) When the present owners were investigating the possibilities of dredging from the lower to the upper ground, 27 churn-drill holes were sunk in the 8,000-foot interval between the two previously prospected areas. The rows of holes were about 750 feet apart, and the holes were spaced 125 feet from each other in the rows. The same drill rig was used in this prospecting that had been formerly used.

---

63 Requa, Lawrence K., Description of the Property and Operations of the Lewiston Dredge, Lewiston, Calif.: Inf. Circ. 6660, Bureau of Mines, 1932, pp. 3-4.

The cost of prospecting can be stated only for this last stage of the work, which was done in 1928. The wage scale was as follows:

	Wages per day
Driller	$6, plus board and transportation.
Panner	6, do.
Fireman	5, no board nor transportation.

The cost was $7.23 per foot. This included the salary of the engineer in charge and the entire cost of a new string of drill pipe - which, however, was not worn out in the prospecting.

The relative merits and applications of shaft-sinking and churn-drilling are the subject of some difference of opinion. Usually, where either method can be applied, shaft-sinking is slower and more costly but more accurate; furthermore, it yields more definite information as to the physical characteristics of the gravel. Any one of these factors may assume such importance as to remove all doubt which method to use. For instance, if the values in a tract of dredging ground are close to the margin of profit by dredging, accuracy of sampling becomes of primary importance, and the cost and speed of sampling have less moment. Conversely, if the sampling is being done for purposes of dredge control on land already held, the cost probably will be the determining factor.

Some ground cannot be sampled accurately by drilling, either because it is loose and coarse, which makes it difficult to recover the gold, or because it is so wet and soft that it is impossible to be reasonably sure of the true volume of sample obtained. Under such conditions shafts should be sunk even if caisson methods are necessary. For a small job churn-drilling would be too costly because of such expenses as first cost of drilling tools and casing, moving into district, and similar overhead or general charges, unless the location is such that equipment can be rented cheaply or unless some contract driller will make a reasonably low bid. Shaft-sinking normally requires very little equipment. Any mining district will furnish miners to sink the shafts, and shaft-sampling is relatively simple and subject to repeated checking as long as the shafts remain open. On the contrary, churn-drilling and churn-drill sampling require the services of an organization whose experience with this particular work and whose reliability and integrity will permit confidence to be placed on the returns.

If it is assumed that the drilling and sampling have been done in the most careful and suitable fashion the significance of the recorded quantities of gold washed from each sample depends wholly on the ensuing computations. First, an assumption must be made as to the volume of gravel represented by each foot of hole drilled; second, it is customary to apply sometimes in the same volume factor, sometimes as a separate calculation in the head office, a correction for losses in dredging. Only wide experience can teach the engineer what these factors should be, as he cannot see what happens a the cutting shoe of the drill, and no satisfactory determination of dredge recovery has ever been made. The drill removes, theoretically, a cylinder of gravel equal in diameter to the outside diameter of the cutting shoe and 1 foot long for each foot drilled. Actually, some gravel may be pushed ahead or aside, or some may run into the hole from outside the casing. In either case, if the deficiency or excess can be detected by observation of the sludge a correction must be applied; this is generally an empirical factor that is accurate only in proportion to the engineer's experience in such work.

Because the outside diameter of the standard cutting shoe for 6-inch casing is 7 1/2 inches the maker of Keystone drills recommends use of 0.3068 square foot to represent the average cross-sectional area of the hole drilled, this being the exact area of a 7 1/2-inch circle. The complete formula reads:

Value of gravel in cents per cubic yard equals weight of gold recovered in milligrams times value of gold in cents per milligram times the conversion factor 27 (cubic feet per cubic yard), divided by the depth drilled in feet times the cross-sectional area of the hole, or briefly,

$$\underline{V} \text{ (cents per yard)} = \frac{\text{weight of gold (mg)} \times 0.06 \times 27}{\text{depth drilled (feet)} \times 0.3068}.$$

The factor 0.3068 is sometimes changed for various reasons. Other common factors are 0.2700 and 0.333, which represent the sectional area in square feet of 7- and 7 13/16-inch holes, respectively. Use of the smaller figure, known as "the Radford factor", raises the estimated value of the gravel about a tenth, whereas the larger figure has the opposite effect.

In the drilling of the area on Trinity River estimates for part of the field were based on the use of the factor 0.27. As it was believed later that these results were erroneously high the adjacent ground drilled subsequently was estimated by the use of the regular Keystone factor 0.3068.[64]

There is a great difference of opinion among engineers regarding interpretation of the results of drill sampling. Some would take virtually no account of sludge measurements. Others insist that the sludge should be measured in calibrated boxes and a correction applied accordingly. Still others make a practice of settling and measuring the mud from the run-off and at the end of drilling make use of the volume of this slime to correct their sludge volumes. Another procedure is suggested by W. W. Avery,[65] who states that the height of core in the pipe should be measured both before and after driving the pipe and a correction made based on the actual as compared with the theoretical rise. The assumption is that the rise of the core when driving the pipe should exactly correspond to the volume of material displaced by the walls of the pipe; if not, an excess or deficiency in the sample is indicated which should be compensated in estimating the values per cubic yard.

The interpretation of placer drilling and the effect of spacing of holes also have been discussed widely. In a recent contribution to the subject[66] the author argued briefly that some maximum spacing, or maximum undivided liability per hole, should be fixed for all drilling which was to be the basis of an examining engineer's report. The ensuing discussion by a large number of engineers brought out a distinct preference for relying on the engineer's education and experience but also the need for a clear statement of the method of making the estimate. It was said that the application of drill and recovery factors could affect the value of a deposit as much as 25 to 50 percent. One New York mining company was said to have had such difficulty in interpreting its engineers' reports because of the different factors used by each that they finally were instructed to report merely the basic figures, such as are contained on drill logs, after which the calculations were made in the home office.

One theory of drill sampling is that the drill holes should be spaced wide apart at first, then checked by subdividing the original network of lines or squares.[67] If the result of all the drilling is different from that of the first set of holes, still further work at closer intervals is desirable until the addition of holes fails to alter the estimate of value by an important amount. Such a procedure must, of course, be planned carefully in ad-

---

64 Requa, Lawrence K., work cited, pp. 4-5.

65 Avery, W. W., Estimating Drill-Hole Data in Placer Prospecting: Eng. and Min. Jour., vol. 129, 1930, pp. 493-495.

66 Rumbold, W. R., The Valuation of Alluvial Deposits: Trans. Inst. Min. and Met., vol. 37, 1928, pp. 437-541.

67 Hutton, H. G., Valuation of Placer Deposits: Min. and Sci. Press, vol. 123, Sept. 10, 1921, pp. 365-368.

vance, and the engineer must use his judgment as to the importance of the possible error and balance this against the known cost and delay of additional drilling.

R. G. Smith believes that average drill results are low because test shafts sunk on drill holes usually give higher returns.[68] He cites 4 sets of 3 to 5 shafts each, which individually indicated values ranging from 96 to 191 percent of the drill-hole returns. By groups, however, the average shaft value was close to 140 percent of the drilling results. This was in ground so firm that the shafts stood 30 to 60 feet deep without timbering. The drill factor 0.30 was used.

At a dredge property in Idaho it was stated recently by the operators that shaft sampling had given results two or three times higher than drilling. The discrepancy was believed to be due to the presence of about a foot of loose sand just above bedrock and to the bedrock being so soft as to hinder drill recovery.

A good discussion of the accuracy of drill sampling of dredge ground is given by C. W. Gardner, a veteran dredge operator.[69] The data contained therein are shown in table 5. Excluding all areas of less than 20 acres and not considering the last four listed, which were sampled by sinking shafts, the average departure from 100-percent recovery is 27 percent plus or minus; the average gain is very nearly equal to the average loss. Gardner states:

> From all the properties above mentioned it is possible to segregate 3,743 acres, to which data given in fairly accurate reports can be applied. This combined area was prospected by means of 1,749 drill holes, or one to every 2.1 acres. The average value per cubic yard obtained by drilling was 15.4 cents and the average dredge recovery 13.55 cents, or 88 percent.

It is apparent that the sampling of dredge ground is far from a precise science. Certainly such large discrepancies would not be expected in sampling low-grade ores, and obviously no refined calculations can be based on such work. On the other hand, it must be remembered that no final proof of the accuracy of placer sampling is ever possible and that dredging involves a metallurgical operation which is subject to many disturbing effects. These facts are discussed further in the chapter on dredging in a subsequent paper.[70]

Geophysical Prospecting

In the last decade geophysical prospecting has been applied to placer mining. Where the bedrock is of homogeneous or uniform magnetic permeability and where considerable magnetite is associated with valuable gold concentrations, magnetometer surveys can outline these leads quite accurately.[71] The instruments used most for this purpose are the Thompson-Thalen, Askania Schmidt field balance, and Hotchkiss Superdip magnetometers, which measure slight changes in the intensity of the earth's magnetic field. Operation of these instruments is not difficult, but correct interpretation of the results requires experience.

Where uniform bedrock conditions do not exist the magnetic work must be supplemented with electrical work to obtain complementary data for use in interpreting the magnetic anomalies. The electrical work gives information on thickness of gravel and depth to bedrock, contour and outline of the underlying bedrock, and other structural features which may be of value in planning the mining operations. It also gives height of the water table.

---

68 Smith, R. G., The Discrepancy Between Drilling and Dredging Results: Eng. and Min. Jour., vol. 112, Nov. 19, 1921, pp. 812-815.

69 Gardner, C. W., Drilling Results and Fredging Returns: Eng. and Min. Jour., vol. 112, 1921, pp. 646-649.

70 Gardner, E. D., and Johnson, C. H., Placer Mining in the Western United States: Part III. - Dredging and Other Forms of Mechanical Handling of Gravel, and Drift Mining: Inf. Circ. 6788, Bureau of Mines, 1934, 81 pp.

71 Laylander, K. C., Magnetometric Surveying as an Aid in Exploring Placer Ground: Eng. and Min. Jour., vol. 121, Feb. 20, 1927, pp. 325-328.

TABLE 5.- Tabulation of dredge recoveries as compared with estimates based on drill sampling[1]

Name, location, or description of tract	Date dredged	Acreage dredged	No. of holes	Spacing of holes	Acres per hole	Average depth of gravel, feet	Value, cents per cubic yard: Estimated from drilling	Value, cents per cubic yard: Recovered by dredging	Percentage recovery	Constant used in estimating	Remarks
Systematically drilled tract; later two modern dredges.		121	50	Rows across, dividing into 8 blocks.	2.4	18	16.8	15.63	93	0.333	Use of factor 0.3068 would have given close agreement.
One 500-foot block in above tract.									49.4		Wide variation of blocks makes the closeness of average remarkable.
Another 500-foot block in above tract.									68.2		
Adjacent portion of above property.		480	130	17 lines across, 400 to 700 feet apart.	3.7	16	11.61	16.44	141.6	.333	Yardage dredged exceeded estimate by 10.1 percent. All blocks gave recoveries ranging between the extremes noted. Similar deposits, same engineer; reason for difference in results unknown.
One block at end of above tract.		2.5							199.2		
Another block in above acreage.		55							182.7		
do.		8.5							104.3		
Property drilled many years ago.		180	[2]33				43.74	19.77	45.7	.3068	
Another part of above property.		118.5	[3]38				29.88	31.55	105.6	.3068	
Small California property (careful tests made on 3 sections).		15.3	27			30.3	3.04	8.7	286	.27	Average estimated value on acreage basis, 5.38 cents per cubic yard; dredge returns, 9.28 cents per cubic yard; recovery, 172.5 percent.
		93	42			22	6.01	10.11	168.2	.27	
		23.5	23			31.5	4.44	6.40	144.2	.27	

1 Tabulated from data by Gardner, C. W., Drilling Results and Dredging Returns: Eng. and Min. Jour., vol. 112, Oct. 22 and 29, 1921, pp. 646-649; 688-692.

2 About 1/2 not in but adjacent.

3 Including 14 on adjacent ground.

TABLE 5.- Tabulation of dredge recoveries as compared with estimates based on drill sampling - Continued

Name, location, or description of tract	Date dredged	Acreage dredged	No. of holes	Spacing of holes	Acres per hole	Average depth of gravel feet	Value, cents per cubic yard: Estimated from drilling	Value, cents per cubic yard: Recovered by dredging	Percentage recovery	Constant used in estimating	Remarks
Pato property, Colombia.		[4]80	[5]19				[4]38	[4]42.5	[4]112	387	Empire drill used. Factor, feet of hole per cubic yard.
Nechi property, Columbia.		[4]60	10	Holes in a line through center of area later dredged.			[4]70.5	[4]35.5	51.3	238	Do.
Chicksan property, Korea.	January-June 1918						14.9	12.25	82	239	Do.
	July-December 1918						11.35	11.30	99.5	239	
	January-June 1919						5.55	9.15	165	239	
	July-December 1919						7.2	10.0	139	239	
	January-June 1920						7.7	12.9	167	239	
Large California property (3 tracts, the third tract having been mined by 3 dredges).		173.5	[6]57		3.2	22.5	6.8	7.82	115	.27	Average (on acreage basis) of drill results, 9.48 cents per cubic yard; dredge returns, 9.12 cents per cubic yard; recovery, 96.2 percent.
		84.0	20		4.2	44.5	5.9	6.7	113	.27	
		183.0	120		1.5	51.8	11.1	9.64	87	.30	
		106.0	41		2.6	60.6	11.2	9.44	84	.30	
		135.0	58		2.3	56.4	11.6	11.30	97	.30	
Alaska Creek		11.76		50 feet apart in lines 300 feet apart.	.25		95.0	96.3	101.4		Used a 6-inch hand drill.
Late report of a California property, one dredge.		157	76		2.1		19.1	10.4	54.4		Values corrected by sludge measurement; factor 0.27 would have given 16.5 cents per cubic yard.
California company, two separate tracts		559.0	560		1.0	22.5	7.58	9.61	126.7	.27	
		420.5	146		2.9	35.9	7.25	8.18	112.9	.27	

4 Approximate.
5 In and adjoining.
6 37 shafts, 20 drill holes.

TABLE 5.- Tabulation of dredge recoveries as compared with estimates based on drill sampling --Continued

Name, location, or description of tract	Date dredged	Acreage dredged	No of holes	Spacing of holes	Acres per hole	Average depth of gravel, feet	Value cents per cubic yard: Estimated from drilling	Value cents per cubic yard: Recovered by dredging	Percentage recovery	Constant used in estimating	Remarks
California property	1918	19.94	11		1.8	32.1	10.39	10.64	102.4		Average on acreage basis, 1918-20; estimated value 10.25 cents per cubic yard; dredge returns 11.39 cents per cubic yard; recovery, 111.1 percent. In 1909 dredge recovery was 58.7 percent; in 1910, 194.4 percent.
	1919	20.90	10		2.1	34.4	9.69	9.22	95.2		
	1920	2043	7		2.9	29.8	10.69	14.34	134.0		
	1908-17	229	173		1.3		10.86	12.34	113.6		
Yosemite Dredging & Mining Co. (Calif)		7	14	In lines across direction of channel.			7.75	6.74	87		
do			66	dc			9		76		Holes ranged from 1 cent per yard or less to 66 cents per cubic yard.
Colorado property; a strip including a line of holes		4 5	14	50 feet apart in a single line		46	17.19	8.76	51.1		Clean, washed sand and gravel; little clay, many boulders.
Colorado property, one dredge.	12 years								165		Medium fine gravel, few boulders.
Do.	1 year	(7)				38		25	290		Do.
Do.			8	8 in a line, cut lengthwise by dredging.			8 6	33	384		Do.
Montana property		300	77	Irregular		40	15.83	13.55	85.6		Most of gold in 3 feet of gravel on bedrock.
California property	Early period	152	23		6.6	35	38.6	16.3	42.2	.27	Medium-size gravel with much clay which hindered washing.

7 1,300,000 cubic yards.

TABLE 5.- Tabulation of dredge recoveries as compared with estimates based on drill sampling - Continued

Name location, or description of tract	Date dredged	Acreage dredged	No. of holes	Spacing of holes	Acres per hole	Average depth of gravel, feet	Value, cents per cubic yard: Estimated from drilling	Value, cents per cubic yard: Recovered by dredging	Percentage recovery	Constant used in estimating	Remarks
Alaskan creek, mined by hydraulicking.	....	(8)	[9]17	....	[0].05	6 to 13	51.6	93.5	180	....	Individual shafts ranged from trace to $1.18 per cubic yard. Gold lay on bedrock covered by boulders.
Idaho dredging property.	....	44	[9]10	(11)	4.4	....	9.9	9.9	100	....	Little water in ground.
California property..	....	140	[9]51	....	2.7	....	15	15.9	106	....	
	....	40	[11]22	....	1.8	....	29	18.2	62.8	....	

8 39,000 square feet.

9 Shafts.

10 Shafts to the acre.

11 Two lines of 5 shafts each at either end of property, 1,500 feet apart, with shafts spaced about 320 feet apart.

12 Mostly shafts.

1776

63

At one proposed hydraulic operation in northern California[72] application of these methods showed that the bedrock slope was quite different from what had been thought and that a buried fault of considerable displacement crossed the property, both facts being of prime importance to the operators. The cost of a combined magnetometric and electrical survey at this property was approximately $2.00 per acre for 600 acres. The results of the survey, made by 2 engineers and 2 assistants in about 30 days, were set forth in a surface topographic and geologic map, a bedrock contour map, and numerous vertical sections across the channel. From these data yardage estimates were made, also recommendations for sinking a few shafts and driving a tunnel on bedrock at points where a maximum of information as to gold value could be obtained.

It must be emphasized that no geophysical method or combination of methods is offered by reputable engineers as a means for determining the commercial value of placer-gold deposits. However, information gained by geophysical methods as to the physical features of the deposit are of great value both in subsequent testing and in developing and operating the property.

Bibliography of Placer Sampling and Estimation

Among the more important treatises dealing wholly or partly with the sampling and estimation of placer deposits are the following:

JANIN, C. W., Gold Dredging in the United States: Bull. 127, Bureau of Mines, 1918, pp. 26-52.

GARDNER, W. H., Drilling for Placer Gold: Keystone Driller Co., Beaver Falls, Pa., about 1923, 196 pp. Includes reprint of article by N. C. Stines, The Prospecting and Valuing of Dredging Ground: Min. and Sci. Press, Feb. 3 and 10, 1906.

WIMMLER, N. L., Placer-Mining Methods and Costs in Alaska: Bull. 259, Bureau of Mines, 1927, pp. 31-40.

JACKSON, C. F., and KNAEBEL, J. B., Sampling and Estimation of Ore Deposits: Bull. 356, Bureau of Mines, 1932, pp. 6-12.

CLASSIFICATION OF MINING METHODS

Placer-mining methods can be classified according to the several methods of excavating and transporting the gravel, or they may be designated to correspond with the various ways of saving the gold. The actual moving of the gravel from place is always the principal concern of the miner, and often the gold-saving is entirely incidental to the working of the deposit. The following classification, therefore, seems the most logical and is the one generally used by placer miners: (1) Hand-shoveling; (2) ground-sluicing; (3) hydraulicking; (4) excavating by teams or power equipment; (5) dredging; (6) drift-mining.

Combinations of methods 1, 2, 3, and 4 may be used, and one method may graduate into another. The methods may be defined as follows:

Hand-shoveling.- Hand-shoveling comprises picking and shoveling surface-placer gravels and washing the material excavated to recover the valuable minerals. The gravel may be washed at the "diggings" or transported by wheelbarrows, pack animals, carts, or trucks to the nearest available water. The general method of excavating varies little, but it may be subdivided according to the method of washing the gravel: (a) Panning; (b) rocking; (c) use of long tom; (d) shoveling into boxes (sluicing); and (e) dry washing.

---

72 Jakosky, J. J., and Wilson, C. H., Use of Geophysics in Placer Mining: Min. Jour., vol. 16, no. 14, Dec. 15, 1932, 3-5 and 29.

<u>Ground-sluicing</u>.- Ground-sluicing consists of moving and washing gravel by water not under a hydraulic head. The action of the water usually is assisted by picking. Boulders may be removed by hand or by derricks. A minor part of the water under relatively low pressure may be used to assist in cutting the bank.

<u>Hydraulicking</u>.- This method of mining utilizes water under pressure from a nozzle for cutting the gravel and sweeping it into sluice boxes, through which it passes to a suitable dumping ground. Additional water not under pressure generally is used to assist in moving the washed material through the sluice boxes.

<u>Excavating by teams or power equipment</u>.- This heading includes mining by teams and scrapers, power scrapers, power shovels, and drag-line excavators but not dredges. The gold may be washed in standard sluice boxes or in more complicated power-driven screening and washing plants.

<u>Dredging</u>.- Dredging properly is confined to mining placer deposits by excavating and washing machinery mounted on boats. The so-called "land dredges" as well as floating washing plants fed by separate dry-land excavators are included under the previous method. The only type of gold dredge operating successfully in this country is the bucket type, which excavates by means of an endless chain of buckets, screens and washes the gravel, recovers the gold in sluices, and discharges the waste material into the water behind the boat.

<u>Drift-mining</u>.- Drift-mining consists of working buried strata of gold-bearing gravels by underground methods. The gravels are mined in much the same manner as flat lode deposits; the excavated material is taken to the surface and washed in sluice boxes or treated in other gold-saving devices. In some instances the milling practice resembles that used for extracting gold from lode ores.

<u>Choice of Methods</u>

The method that involves the lowest total cost is preferable for mining a given deposit. The method chosen, however, must be compatible with existing conditions; frequently only a relatively high-cost method will be practicable. Usually natural conditions restrict the selection of a method. The most important natural factors are the quantity and pressure of water available; the slope or grade of surface and bedrock; the depth, extent, and value per yard of the deposit; the character of gravel and bedrock; and the thickness of the overburden.

For profitable exploitation of a mine the gold recovered must be of greater value than the cost of mining it. Although in the past extremely rich placer ground has been found and worked, present known surface placer deposits are of relatively low grade. For profitable operation, relatively large tonnages must be handled per man-shift; too much stress cannot be laid upon this item. All phases of the mining must be correlated, and delays must be reduced to a minimum. Complications in any part of the set-up are to be avoided whenever possible in order that no "bottle necks" may exist. The mine and plant must be so designed and laid out that operations can be maintained at full capacity throughout the season, especially where power excavators or mechanical washing plants are used. Often the difference between expected capacity and actual capacity has been considerable and has meant the difference between profit and loss.

HAND-SHOVELING

The first placer mining in the West was done by hand methods. As the rich surface deposits became exhausted other methods were substituted. Under normal economic conditions the hand-shoveling method is applied mainly to deposits that are too small to justify the capital expenditure necessary for working them by other methods. When work is scarce, however, any available gravel that will give a man a bare living may be worked in this manner. Because very little capital is required, men of small means can mine on their own account by hand-shoveling methods; in times of general unemployment, many such men go into the placer fields and search for and work surface gravels. This type of mining commonly is called "sniping."

During the summer of 1932 thousands of men, most of them inexperienced, were mining in the placer districts of the West by hand-shoveling methods. They seldom earned a "going wage" but usually got enough gold to buy food. The average earnings of experienced men were estimated to be less than $1 per day and of experienced and inexperienced together less than $0.50. Occasionally a rich spot left by oldtimers would be found and as much as $100 might be cleaned up in a few days. These rare experiences, however, were offset by many days spent searching for a place to work or by wasted effort in digging to bedrock only to find it already cleaned or barren. During the autumn of 1931 many old bars in the rivers of the Pacific coast were exposed for the first time in years on account of the extremely low water of that season. Many snipers worked profitably on these bars until the fall rains raised the water level.

Snipers look mainly for unworked ground in or around old workings, or they extend old workings that have been discontinued as being unprofitable. They reclean disintegrated bedrock in old workings or clean out crevices in bedrock in river channels. They work old bars in the rivers where reconcentration may have occurred since the early days of placering. Most of the ground worked by snipers, however, is ground that has been left by oldtimers as being unprofitable.

In dry sections of the country old channels, dry creek beds, or other gold-bearing deposits are worked, and the gravels are cleaned with dry washers.

Most of the ground in the Western States known to contain placer deposits is owned privately or held under mining locations. A newcomer into a field generally has to obtain permission from the proprietor of the land before he can begin placer mining. Sometimes a royalty is charged; at other times hand mining is permitted free, the owners of the gravel doubtless considering that the prospecting value of such work equals the small amount of gold taken from the ground.

The water of most western streams has been appropriated for irrigation or power purposes. Placer miners should consider the previous rights of others in a stream before diverting any water for placer mining. The antidebris law of California prohibits uncontrolled discharge of placer tailings into any tributaries of the Sacramento River.

Panning

Panning is both the simplest and most laborious method of washing gold from placer gravel. With a little experience almost any one can recover most of the gold in a pan of ordinary placer dirt; an experienced man can wash about 10 pans per hour. As a man can dig gravel with a pick and shovel many times faster than he can pan it only the highest-grade gravel is washed. The top dirt usually is thrown aside, and a few inches of gravel on bedrock or material scraped from crevices is panned.

The standard gold pan used in the Western States is made of stiff sheet iron; it is 16 inches in diameter at the top and 2 1/2 inches deep. The rim is flared outward at an angle of about 50° from the vertical. Smaller pans are used for testing. Frying pans or other cooking utensils may also be used for washing out gold. Before any kind of a vessel is used for panning it should be cleaned thoroughly; all grease should be burned out. New pans generally are greasy and should be heated over a fire until all grease is burned off. A rusty pan if clean can be used satisfactorily; the roughness due to the pitting of the rust may be of assistance in holding back the gold.

The pan usually is filled even with the top, or slightly rounded, depending somewhat upon the nature of the material being washed and the personal preference of the panner; it is then submerged in the water. Still water 6 inches to 1 foot deep is best. While under water the contents of the pan are kneaded with both hands until all clay is dissolved, and lumps of dirt are thoroughly broken. The stones and pebbles are picked out. Then the pan is held flat and shaken under water to permit the gold to settle to the bottom. The pan is then tilted and raised quickly so that some of the lighter top material is washed off. This operation is repeated, occasionally shaking the pan under water or with water in it until only the gold and the heavy minerals are left; this material concentrates at the edge of the bottom of the pan. Care must be taken that none of the gold climbs to the edge of the pan or gets on top of the dirt.

Nuggets and coarse colors can now be picked out readily. Cleaning the black sand from the finer gold is more difficult but can be carried nearly or entirely to completion by careful manipulation of the pan as described above, always watching carefully to see that none of the colors are climbing toward the edge. This part of the operation usually is done over another pan or in a tub so that if any gold is lost it can be recovered by repanning.

The concentrates can be dried and the black sands removed by a magnet or by gentle blowing on a smooth flat surface. If there is an excessive quantity of black sand the gold usually is amalgamated by putting about a teaspoonful of mercury in the pan. In sampling work great care is necessary that no fine colors are lost, but in mining by panning the extra time needed to make sure that no fine colors escape probably would not be justified. The loss of 50 or 100 fine colors in a day might not amount to more than 1 cent.

A standard 16-inch pan level full of dry bank gravel contains, on an average, 22 pounds. The quantity will vary with the moisture content and nature of the gravel. By allowing 3,300 pounds to the cubic yard (in place), 150 pans would be equivalent to 1 cubic yard. The senior author has been accustomed to using this factor in sampling placer deposits. At 3,600 pounds to the cubic yard and 22 pounds to the pan, 164 pans would be required to wash a cubic yard of gravel. At 10 pans per hour a man could pan about 1/2 cubic yard in a day. With exceptionally clean gravel a man will sometimes pan as much as a cubic yard in a day.

Rocking

More gravel can be handled per man-day by rocking, or cradling as it is sometimes called, than by panning. Moreover, the manual labor of washing a cubic yard is less. The same method of excavating the gravel is used whether it is panned or rocked. The rocker, like the pan, is used extensively not only in small-scale placer work but also in sampling and for washing sluice concentrates and material cleaned by hand from bedrock in other placer operations.

One to three cubic yards, bank measure, can be dug and washed in a rocker per man-shift, depending upon the distance the gravel or water has to be carried, the character of the gravel, and the size of the rocker. Rockers usually are homemade and have a variety of designs. A favorite design in the Western States consists essentially of a combination washing box and screen, a canvas or carpet apron under the screen, a short sluice with two or more riffles, and rockers under the sluice. The bottom of the washing box consists of sheet metal with holes about one half inch in diameter punched in it. A rocker in use at Greaterville, Ariz., was 3 feet 4 inches long and 1 foot 9 inches wide on the inside and had a slope of 5 inches. The screen box was 6 inches deep and 20 inches square inside and had a bottom of sheet iron with 1/4- to 1/2-inch holes punched about 2 inches apart. The baffle was 28 inches long and consisted of a piece of canvas. A single riffle 3/4 inch high was used at the end of the rocker. Figure 3 is a drawing of a prospector's rocker made by W. B. Young of Tucson, Ariz. The bottom of a rocker should be made of a single wide board, if one can be obtained, and planed smooth. This will greatly facilitate clean-ups. The cost of building rockers ranges from $5 to $15, depending mainly upon the cost of lumber.

After being dampened the gravel is placed in the box 1 or 2 shovelfuls at a time. Water is then poured on the gravel while the rocker is swayed back and forth. The water usually is dipped up in a long-handled dipper made by nailing a tin can to the end of a stick. A small stream from a pipe or hose may be used if available. The gravel is washed clea n in the box and the oversize inspected for nuggets and dumped out. The undersize goes over the apron, where most of the gold is caught. Care should be taken that too much water is not poured on at one time, as some of the gold may be flushed out. The riffles stop any gold that gets over the apron. In regular mining work the rocker is cleaned up after every 2 or 3 hours, or oftener when rich ground is worked, if gold begins to show on the apron or in the riffles. In cleaning up after a run, water is poured through while the washer is gently rocked; the top sand and dirt are washed away. Then the apron is dumped into a pan. The material back of the riffles in the sluice is taken up by a flat scoop, placed at the head of the sluice, and washed down gently once or twice with clear water. The gold remains behind on the boards, whence it is scraped up and put into the pan with the concentrate from the apron. The few colors left in the sluice are caught with the next run. The concentrate is cleaned in the pan.

With skillful manipulation of the rocker and a careful clean-up nearly all the gold is recovered. Violent rocking is avoided so that gold will not splash out of the apron or over the riffles. The sand behind the riffles should be stirred occasionally, if it shows a tendency to pack hard, to prevent loss of gold. If the gravel is very clayey it may be necessary to soak it for some hours in a tub of water before rocking it.

When water is scarce two small reservoirs are constructed, one in front and the other in the rear of the rocker. The reservoir at the front serves as a settling basin; the overflow goes to the one at the rear where the water is used over again.

Power rockers.- The capacity of rockers may be increased by using power drives. The use of such a machine was illustrated by the operation of George Graves in the Lynn district, Eureka County, Nev., during the summer of 1932. The rocker was 49 inches long, 27 inches wide at the top, and 21 inches wide at the bottom. It was 24 inches high in front and 21 inches at the rear. The screen had 5/8-inch round holes. The gold was caught on three aprons of canvas and wood. Riffles of 1/2- by 1/4-inch wooden strips were used on the aprons. The undersize from the screen passed over each apron in turn. Nearly all the gold was caught on the first apron. The slope of the aprons was 3 inches to the foot.

The device was rocked by an eccentric arm at the rate of forty 6-inch strokes per minute. The capacity of the machine with two men working was 1 cubic yard per hour. Where gravel was free of clay the capacity was said to be as great as 3 cubic yards per hour. The cost of the rocker and the engine for driving it was $160. At $4 per 8-hour shift and 1 cubic yard per hour the labor cost of washing the gravel would be $1 per cubic yard.

A number of small machines patterned more or less after the power rocker are on the market. They usually are built of iron or steel and driven by small gasoline engines. Although of various designs they generally consist of a trommel or a shaking screen to remove coarse material, a short shaking sluice to save the gold, and a pump to circulate the water. Some of them contain a settling tank from which the solids are removed by a rake or drag. These machines have an advertised capacity of 1/2 to 2 1/2 cubic yards per hour and cost $225 to $700. No operating data are available.

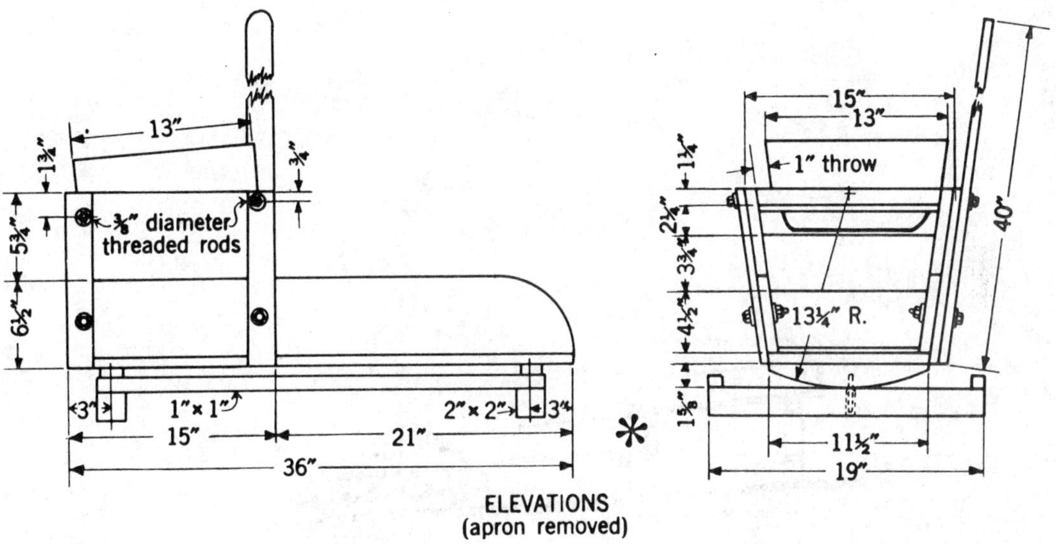

ROCKER CAN BE POWERED BY GASOLINE ENGINE.

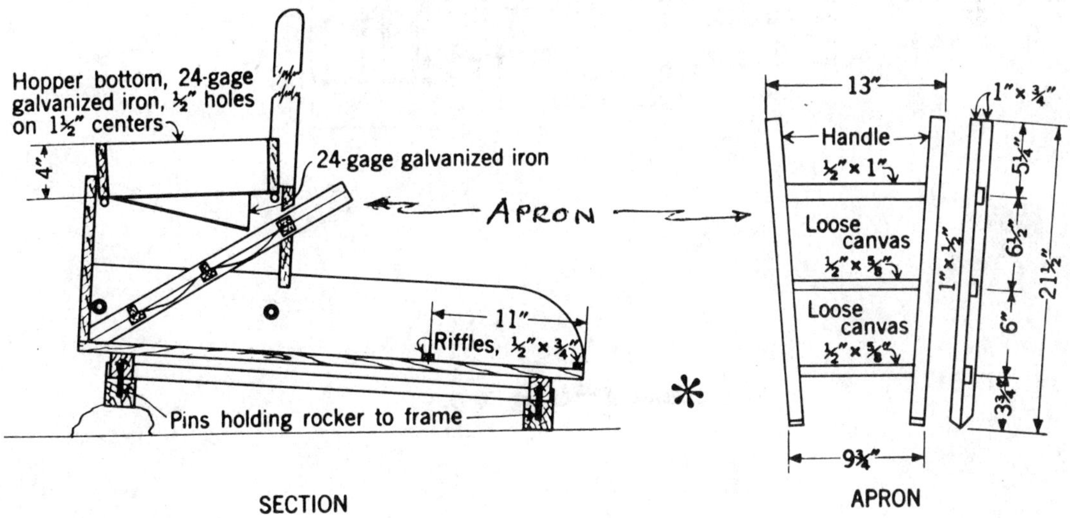

Figure 3.—Prospector's rocker.

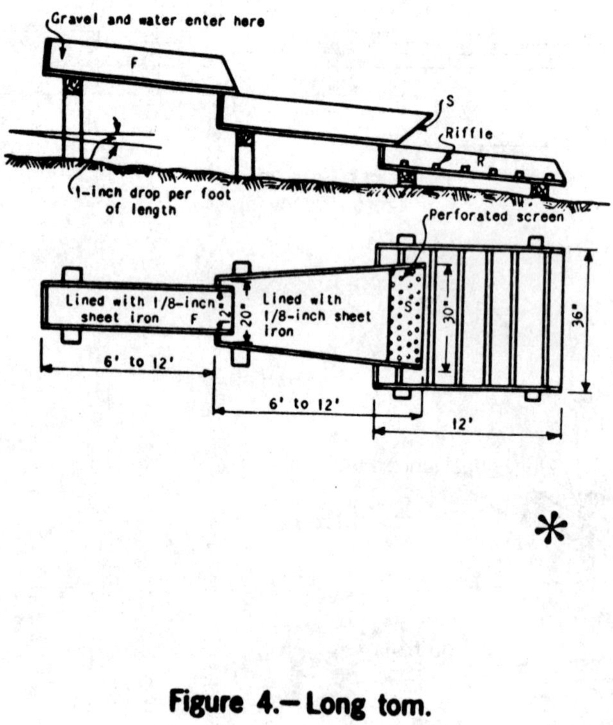

*

Figure 4.—Long tom.

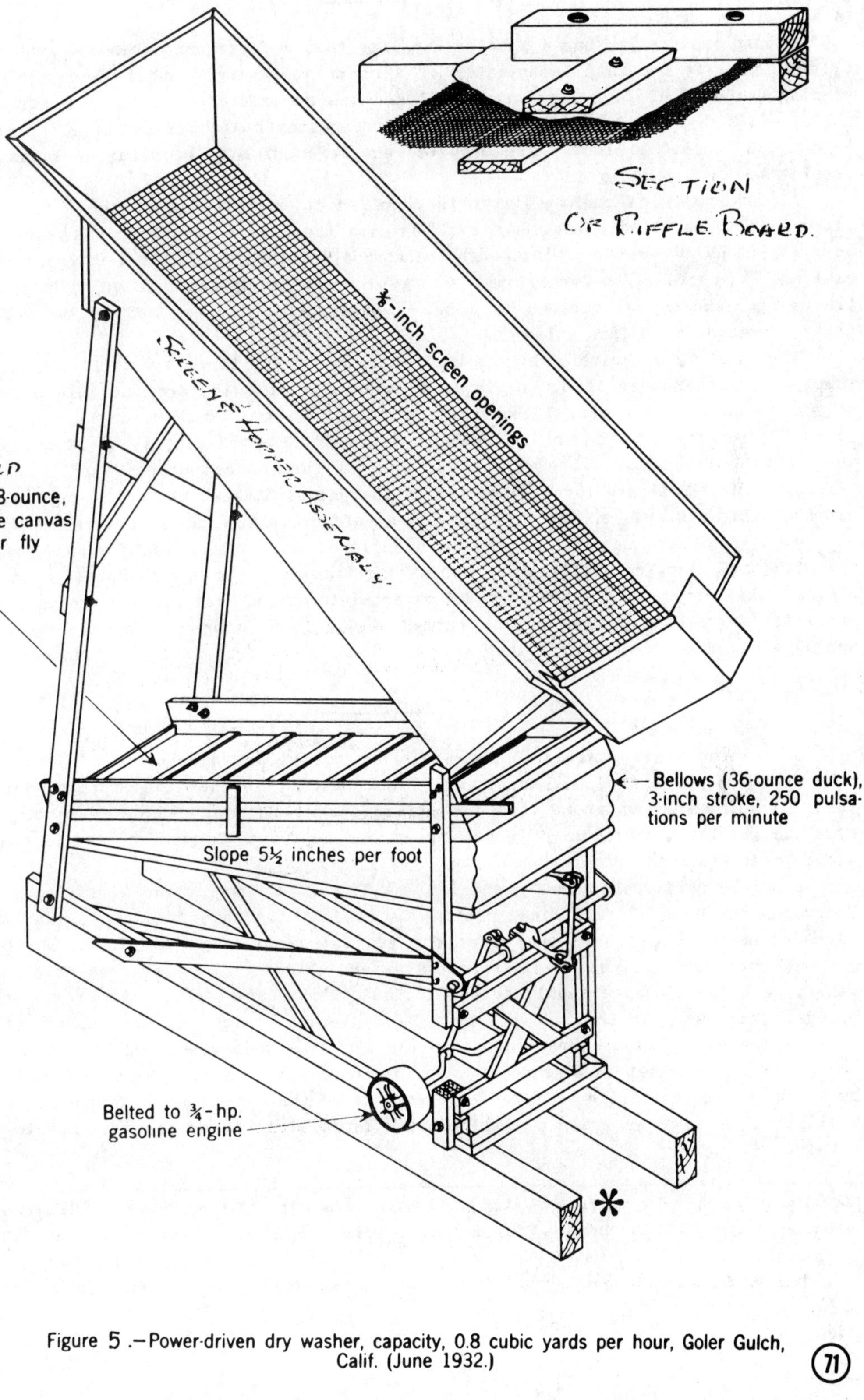

Figure 5.—Power-driven dry washer, capacity, 0.8 cubic yards per hour, Goler Gulch, Calif. (June 1932.)

Long Tom

A long tom usually has a greater capacity than a rocker and does not require the labor of rocking. It consists essentially of a short receiving launder (see fig. 4), an open washing box 6 to 12 feet long with the lower end a perforated plate or a screen set at an angle, and a short sluice with riffles. The component boxes are set on slopes ranging from 1 to 1 1/2 inches per foot. The drop between boxes aids in breaking up lumps of clay and freeing the contained gold.

A good supply of running water is required to operate a long tom successfully. The water is introduced into the receiving box with the gravel, and both pass into the washing box. The sands and water pass through the screen, which has about 1/2-inch openings into the sluice. The oversize is forked out. The gold is caught by the riffles. The riffle concentrates are removed and cleaned in a pan. Quicksilver may be used in the riffles if the gravel contains much fine gold.

The quantity of gravel that can be treated per day will vary with its nature, the water supply, and the number of men employed to shovel it into the tom and fork out the stones. Wilson[73] states that 2 men, 1 shoveling into the tom and 1 working on it, can wash 6 cubic yards of ordinary gravel, or 3 to 4 cubic yards of cemented gravel, in 10 hours. At times the tom is operated by 4 men, 2 shoveling in, 1 forking out stones, and 1 shoveling fine tailings away. Toms are rarely used now in the United States; where running water and grade are available a simple sluice is generally as effective and requires less labor.

A modified form of long tom has been employed for washing beach sands at Nome, Lituya Bay, Yakataga, and Kodiak, Alaska. For saving fine gold the box is set at a high gradient, 3 to 4 inches per foot, and the screened material is passed over riffles and amalgam plates.[74] The water for sluicing generally is bailed with a large dipper. The average duty per man per 10 hours for long-tom or rocker work on beach operations is 3 to 5 cubic yards.

Surf Washers

Surf washers are similar to long toms but are wider and shorter.[75] They can be used only when the surf is of proper height. They are set so that the incoming surf rushes up the sluices, washes material from the screen box or hopper, and on retreating carries it over the riffles and plates. One man can attend to two surf washers, and in one instance 8 cubic yards was handled per 10 hours.

A simple surf washer used about 1902 at Topkuk, Alaska,[76] was a riffled sluice 3 to 4 feet wide and 8 to 10 feet long, set on the sand at the water's edge so that the incoming waves washed through it to the upper end yet retreated below the lower end. The sluice was made of boards nailed to sills at either end which could be weighed down with rocks or otherwise. The sides were only 4 or 5 inches high. The riffles, similar to those used on dredges, were made in sections of about 1- by 1-inch strips and were spaced about an inch apart. The end sections were transverse riffles; the center section was longitudinal. The box preferably was set on a grade of 8 to 10 inches per 12 feet. Best results were obtained by using quicksilver in the riffles. It was stated that when the surf was strong the washer would treat as much as two men could shovel, but that at other times it had to be fed very slowly.

---

73 Wilson, E. B., Hydraulic and Placer Mining: John Wiley & Sons (Inc.), New York, 3d ed., 1918, 425 pp.

74 Wimmler, Norman L., Placer Mining Methods and Costs in Alaska: Bull. 259, Bureau of Mines, 1927, 236 pp.

75 Wimmler, Norman L., work cited.

76 Elffner, A. E., Beach Mining with a Surf Washer: Min. and Sci. Press, vol. 86, June 6, 1903, p. 364.

Dry Washers

Dry washers have been used for many years in the Southwest where water is scarce, especially in New Mexico where probably 3 or 4 million dollars in gold has been produced during the last century by dry-washing. A small steady production by dry washers still comes from the Cerrillos, Golden, and Hillsboro districts each year. In years when other employment is scarce the production increases. During 1931 and 1932 a considerable number of men also used dry washers in Nevada, southern California, and Arizona.

Gravel to be treated successfully by dry washers must be completely dry and disintegrated. With dry washers operations must be stopped after rain storms until the ground dries out again. The only successful dry washing up to 1932 has been on a small scale. Plants with mechanical excavators and extensive power-driven dry-washing machinery have been tried, but in the United States none has been commercially successful, partly because in large-scale work the gravel is dug faster than the sun can dry it out. Even in very dry climates the gravel is slightly damp below the surface and must be dried before it can be treated in a dry washer. Spreading the material to sun-dry or putting it through driers adds to the cost of mining. In small-scale work the gravel dries out as fast as it can be treated.

Individual workers select the material they treat with regard to both dryness and probable gold content; it is difficult to do this on a large scale with hired labor. In large-scale work, particularly with mechanical excavation, the cost of sizing the material is an important item; clay and cemented gravel introduce further difficulties.

Dry washers usually are run by hand and have about the same capacity per day as rockers of a corresponding size, but the work of operating the dry machine is much harder. When the gold-bearing material is completely dry and disintegrated panning tests of the tailing show that a good saving can be made, except perhaps with extremely fine or flaky gold. Completely disintegrated material, however, seldom is obtained. The tops of clay streaks in the gravel are likely to be richer in gold than the gravel itself. Clay or cemented gravel seldom can be broken up sufficiently by hand to free all the gold without the use of some form of pulverizer. In a dry machine all gold included in a lump of waste passes out of the machine. As water usually breaks up all the gravel and separates the gold from the other material a better saving generally can be effected by rocking or in sluice boxes than in a dry machine.

The basic principle of the dry washer is separation of the gold from the sand by pulsations of air through a porous medium. The screened gravel passes down an inclined riffle box with cross riffles; the bottom of the box consists of canvas or some other fabric. Under the riffle box is a bellows by which air in short, strong puffs is blown through the canvas. This gives a combined shaking and classifying action to the material; the gold gravitates to the canvas where it is held by the riffles, while the waste material passes out of the mahine.

The gravel is shoveled into a box holding a few shovelfuls at the head of th e washer whence it runs by gravity through the machine. A screen with about 1/2-inch openings is used over the box; all stones over about 1 inch in diameter generally are discarded in mining. one man working alone fills the box, then turns a crank which runs the bellows until the gravel is run through; the process is then repeated. With two men working, one shovels and the other turns the crank. One man can treat 1/2 to 1 yard per day with a hand-operated washer where the gravel is handy to the machine. Some dry washers are run by a small gasoline engine which saves the labor of one man; the capacity of such machines is considerably greater than that of hand-operated machines.

In cleaning up, the material back of the riffles usually is dumped into a pan and washed out in water. If water is very scarce the accumulated material from the riffles may be run through the machine a second time and then cleaned further by blowing away the lighter grains of sand in a pan.

Dry washers usually are handmade and have a large number of designs and sizes. Figure 5 is an isometric drawing of a machine built and run by E. C. Spurr on Goler Gulch near Randsburg, Calif., in June 1932. The bellows of the machine is made of 36-ounce duck and the bottom of the riffle box of 8-ounce single-weave canvas. Spurr has found by long experience that the 8-ounce material is the best for the purpose. Silk or rayon permits a good extraction of gold, but too much dust goes through into the bellows. Heavier canvas is too tight for good separation. Copper-wire fly screen is used under the canvas. The riffle box is 11 inches wide and 40 inches long and contains 6 riffles. The type of riffle is shown in figure 5. The slope of the riffle box is 5 1/2 inches to the foot. The gravel and sand are shoveled onto a screen with 3/8-inch square openings at the top of the washer. The machine is run by a 3/4-hp. gasoline engine. The bellows is operated at 250 pulsations per minute; the stroke is 3 inches. The capacity of the machine was 1 cubic yard in 1 1/4 hours or 0.8 yard per hour. This probably would correspond to 1 1/2 or 2 cubic yards, bank measure, on account of the plus 1-inch material having been thrown out in mining. The gravel had been collected in an old drift mine and carried out in buckets; 3 hours were spent by one man in getting out the gravel.

In cleaning up after running a cubic yard through the Spurr washer the riffle box was lifted out and turned over on a large, flat, baking tin. The concentrate from the upper three riffles was first panned and the gold removed; both coarse and fine gold were saved, the total being worth about 75 cents. The lower four riffles had one good-sized color and a few small ones. Nearly all the gold was caught in the upper riffles; this indicated that a fair saving was being made of the free gold.

Hand-operated machines, however, are usually much smaller and have the riffle box set at a steeper angle.

Shoveling-into-Boxes

Mining.- Shoveling-into-boxes is a small-scale placer-mining method in which the gravel is loosened by picking and is shoveled by hand into sluice boxes. It is particularly adapted to small, shallow, moderately rich deposits where only a little water is available or where the grade is insufficient to allow room for disposing of the tailings unless the gravel is elevated. If the water supply is adequate the ground-sluicing method (described later) should be considered, and if the deposit is large enough to justify the required capital investment some other method such as hydraulicking or dredging would be preferable.

The method of shoveling-into-boxes has its chief application in the working of small deposits by men with little capital but a willingness to work hard.

The quantity of water available will influence the scale of operations and the size of sluice used. A minimum flow of 15 to 20 miner's inches (170 to 225 gallons per minute) is required for a 12-inch sluice box with a steep grade. Smaller flows than this can be utilized by reservoiring the water and using an augmented flow. A common practice followed where the quantity of water is limited is to place a grizzly or screen over the sluice and thus increase the duty of the water. A head box with sloping sides and a grizzly near the bottom may be used for receiving the gravel. In this case the oversize is forked out. In one Nevada operation the gravel first was run through a trommel to wash it and screen out all coarse material. Puddling boxes may be used if the gravel contains much clay. If the ground is of good grade it may be practicable to pump water for the sluice; the feasibility of obtaining a gravity flow, however, should first be investigated, as the mining expense due to pumping may be more than the cost of a ditch distributed over the yardage moved.

Water usually is conducted in a ditch to the sluice which is set up in the most advantageous position to begin work. Boxes are 8 to 24 inches wide and range from 3 to as many as may be required to transport the tailings to a suitable dumping ground; they are laid on sills at approximately the mean elevation of the surface of the deposit. If enough water is available 6 or 8 inches per 12 feet is a good grade with most types of gravel. If water is scarce or the gravel angular a steeper grade will be desirable; the grade should be uniform. The best type of riffle depends on the nature of the gravel and the gold rather than on the method of extraction. An end shake was given to the box at a mine in Arizona to increase its capacity. Water was hauled for washing. Riffles are discussed in a subsequent paper.[77]

The cost of boxes depends chiefly on the cost of lumber delivered to the property, which may range from $15 to $60 per 1,000 board-feet. A 12-foot box, 12 inches wide, conructed of three 1- by 12-inch boards with three sill frames of 2 by 4 lumber, requires about 45 board-feet of lumber. If clear lumber costs $35 per 1,000 board-feet the cost of a set of five boxes would be about $8.00. Material for riffles may cost little or nothing or as much as $1 per box.

After the boxes are set, shoveling begins at an advantageous point. All material of a size that will run through the sluice is shoveled in, and the oversize is thrown to one side. Boulders from the first cut should be stacked outside the pit, on barren ground if possible. The width of a cut usually is limited to the distance a man can shovel in one operation. When shoveling from any distance it is best to set boards above and on the opposite side of the box; these increase the efficiency of the shovelers. The greatest height a man can shovel into a box is 7 to 8 feet; above 5 or 6 feet the efficiency of the shovelers is reduced markedly. If over 3 or 4 feet in depth the gravel usually is excavated in benches to facilitate digging and at the same time to permit the upper layers of gravel to be raised a minimum height. Where the gravel is shallow wheelbarrows may be used. At a Nevada mine the gravel was shoveled onto a conveyor belt which discharged into a trommel. The oversize was discarded, and the undersize ran through a sluice. Where two or more men are working in the same cut the heights of succeeding benches are governed by the character of the material being dug and the distance the gravel has to be lifted. Work may begin at one edge of the deposit and proceed across it by regular cuts. The sluice may be maintained on the surface of the unworked ground or on bents on the opposite side of the cut. After the first cut the boulders are thrown onto the cleaned-up bedrock. Where cuts are run on both sides of the sluice the boxes are supported on bents as the ground underneath them is dug out. At other places the boxes may be set on bedrock and all dirt shoveled into the head of the sluice from short transverse cuts at the upper end of the pit. Work usually begins at the lower end of a deposit so that bedrock may be kept drained. The length and order of making the cuts will depend upon local conditions.

The method of working a deposit near Oroville, Wash., is shown in figure 6, A, and at Blewitt, Wash., in figure 6, B.

Experienced miners work out the ground in regular cuts and in an orderly fashion. Enough faces are provided so that shovelers will not interfere with one another. Provision is made to keep bedrock drained, and boulders and stumps are moved a minimum number of times. Cuts are taken of such a width and length as to make the shoveling as easy as possible. The boxes are kept at such a height that the minimum lift of the gravel is required consonant with enough fall for the gravel to run through the boxes with the water available and to allow dump room at the tail end of the sluice. Leaks in the sluice are stopped promptly, and shoveling is done in such a manner that the sluice does not become clogged or water splashed out. Water in the pit hampers shoveling.

Several times as much gravel can be washed per man-shift by this method as by the use of rockers or long toms. Under the best conditions a good workman can shovel 15 cubic yards

---

77 Gardner, E. D., and Johnson, C. H., Placer Mining in the Western United States: Part II. - Hydraulicking, Treatment of Placer Concentrates and Marketing of Gold: Inf. Circ. 6787, Bureau of Mines, 1934.

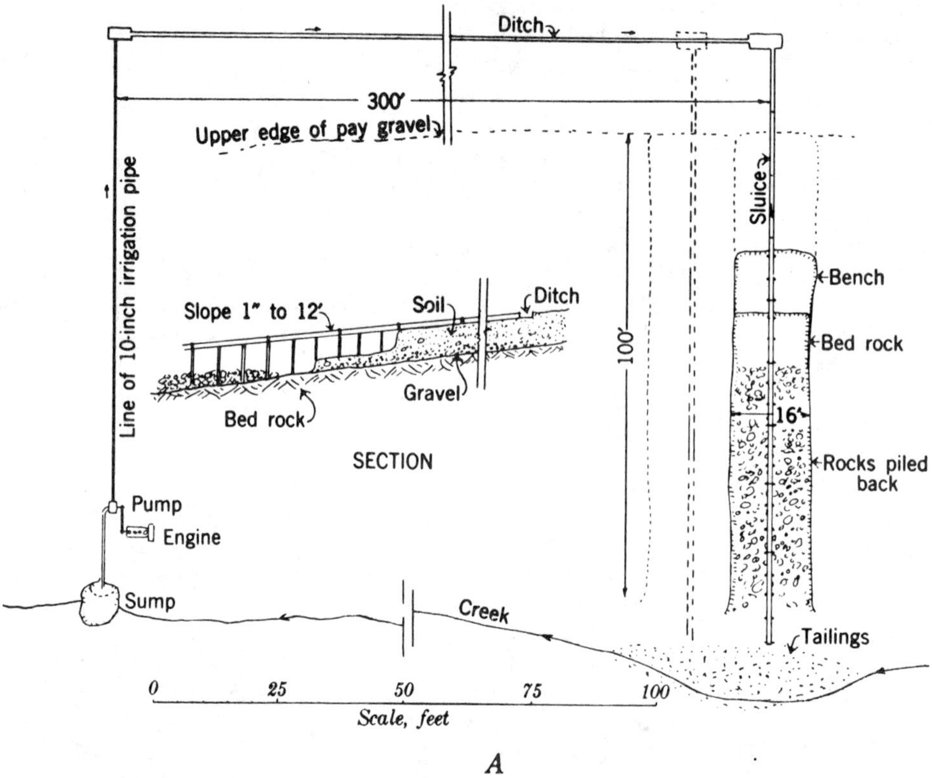

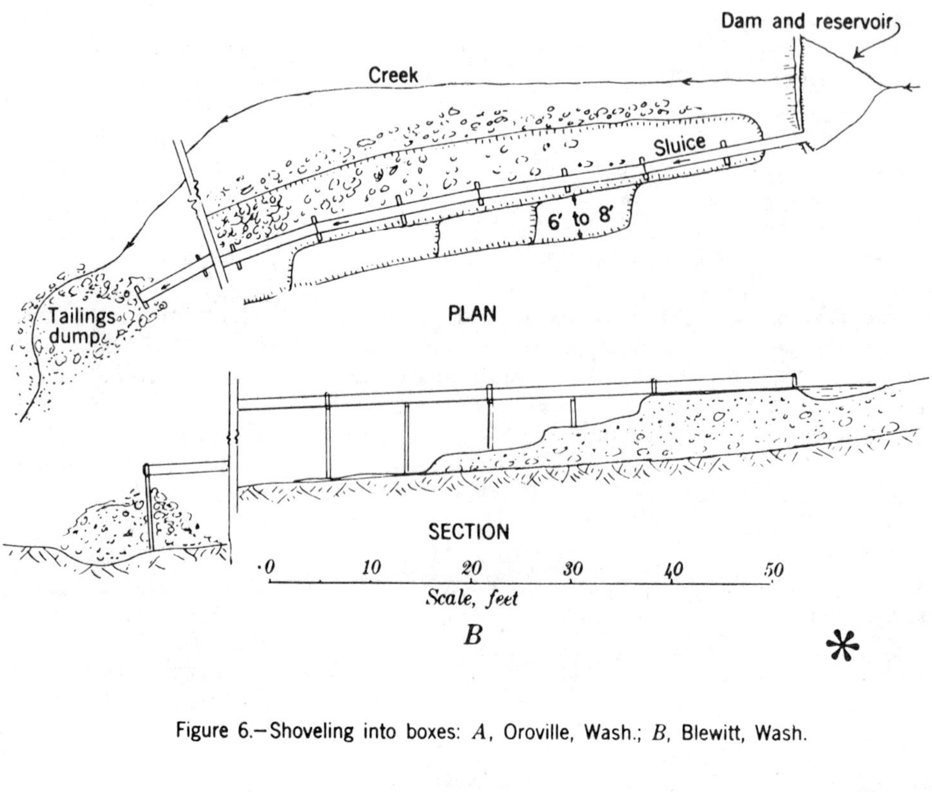

*

Figure 6.–Shoveling into boxes: *A*, Oroville, Wash.; *B*, Blewitt, Wash.

of gravel per shift into boxes 3 or 4 feet above bedrock. Where the gravel is tight and coarse and the bedrock rough, he may be able to shovel only a yard a day. The time necessary to construct, set up, and clean the sluice boxes and to clean bedrock reduces the average performance of men shoveling into boxes to about 4 yards per shift.

Cleaning bedrock.- When a sufficient area of bedrock is exposed it is cleaned up carefully. If the bedrock is easy to clean this may be done as the work progresses up a cut. In such ground if the gravel is shoveled in benches by two or more men the man who takes up the lower layer of gravel may clean the bedrock as he goes along. If the bottom is difficult to clean a full cut may be taken before the bedrock is given the final clean-up.

Cleaning bedrock is perhaps the most laborious and painstaking task of the placer miner; his success or failure often depends on the thoroughness of this operation.

After all loose gravel is shoveled into the boxes the bedrock should, if possible, be taken up with pick and shovel to a depth of 6 inches or more and likewise put into the sluice. Any large rocks near bedrock should be washed off before they are thrown aside, because clay adhering to them may contain gold. If the bedrock is decomposed or soft, or if there is a false bedrock, this process will recover practically all the gold.

If bedrock is hard, cleaning it is more difficult, as some material is bound to be left on the bedrock in shoveling, and the crevices in bedrock will contain gold.

If water can be taken from the sluice or ditch through a hose under low pressure all loose material missed in shoveling will be washed down to a hollow where it will be collected and placed into the boxes or in a rocker. The bedrock then is scraped and swept, and the crevices are cleaned out; finally it may be washed a second time.

If it is not convenient to use water the bedrock may be cleaned up dry. The bedrock is swept, scraped, then swept a second time; the crevices are cleaned out as before. The sweepings and scrapings, together with the material from the crevices, are washed either in the sluice or in pans or rockers. Large colors or nuggets are picked up as found.

The tools used for the hand work are picks and shovels, brooms, small stiff brushes, perhaps a sponge to dry out hollows, and hand scrapers of various sizes and shapes, ranging from sections of old frying pans to knives, spoons, or bent wire for cleaning narrow crevices.

In small operations, such as the typical 3- or 4-man ground-shoveling venture, cleaning bedrock is carried to greater extremes than in larger mines where labor is hired, as the incentive to careful work is greater and the yield necessary to encourage further cleaning is less. Skilled and careful "crevicers" have been known to make fair wages on bedrock abandoned by organized mining companies. The miner should therefore be sure his cleaning operations have been carried to the point where he is no longer getting enough gold to recompense his efforts.

The concentrates from the sluice usually are panned or washed in a rocker for the final separation of the gold.

Current Practices

During 1931 and 1932 hundreds of men were mining by hand-shoveling methods in the West. The miners worked in crews of 1, 2, 3, or 4 men. In a few instances as much as $3 per man was made, but the average earnings were well under $1 per day. Most of these men were inexperienced and did not work to the best advantage. Boulders would be rolled to one side and sometimes handled 3 or 4 times before the work had advanced beyond them. Again, the ground would be cut up in uneven forms and part of the gravel shoveled twice to get it into the boxes. Boxes were not always well made, and considerable gold would be lost through leaks or cracks.

Table 6 shows seven representative mines worked by or under the direction of experienced placer miners. Where the men were inexperienced the yardage handled per shift was much less. The cubic yards per man-shift and the estimated cost per cubic yard with labor at $3 per shift are included in the table.

TABLE 6.- Representative examples of shoveling into boxes

	Oroville	Jones	Peshastin	Minnick	Blackhawk	Greetan	DeWitt
Operator of mine		J. B. Jones				O. F. Greetan	H. DeWitt.
Address	Oroville, Calif.	Oroville, Wash.	Blewett, Wash.	Blewett, Wash.	Blackhawk, Colo.	Bearmouth, Mont.	Wilcox, Ariz.
Location of mine	Feather River	Mary Ann Creek	Peshastin Creek	Peshastin Creek	North Clear Creek	Bear Creek	Gold Gulch.
Number of men working	2	4	4	2	2	2	1.
Hours per shift	9	9	9	8	8	8	8.
Depth of gravel ... feet	2	[1]6	Tight	Tight	Loose	Medium	Tight.
Character of gravel	Loose top dirt	Tight	6	2 1/2	4 1/2	6	
Percentage of boulders over 6 inches in diameter in gravel	0	30	20	50	10	5	
Character of bedrock	None	Clay	Even	Very rough	None	Even	Clay.
Width of sluice boxes ... inches	8	8	12	12	10	8	18.
Grade ... in. per ft.	1	1	7/16	1/2		6/10	
Length of each box ... feet	12	12	16	12	12	10	22.
Number of boxes	2	8		3	3	6	1.
Type of riffles	Hungarian	Steel matting over coco matting.	Hungarian, pole	Wire screen over Brussels carpet.	Wire screen over carpet.	Wooden, cross	Wooden, cross.
Size of riffles ... inches	1 by 1 1/4		Hungarian, 3/16 by 1 1/4; pole, 3.			1 by 1	1 by 1
Spacing of riffles ... do.	1		1			4	1.
Total length of riffles ... feet	12	8	64	12	12	28	18.
Height water pumped ... do.	10	37	0	0	0	0	0.
Percentage of material rehandled	5	10	0	50	0	0	200.
Handled per day ... cu. yd.	[2]20 to [3]30	25	20	2	6 1/2	9 1/2	1.
Handled per man-shift ... do.	[2]10 to [3]15	6 1/4	5	1	3 1/4	4 3/4	1.
Gasoline used per cubic yard ... gal.	3	6	0	0	0	0	1.5.
Estimated cost of labor and gasoline per cubic yard[4]	$0.36 to $0.26	$0.60	$0.60	$3.00	$0.92	$0.63	$3.30.
Remarks	(5)	Good workmen	Good workmen	(6)		(7)	

1 Top 2 feet consists of soil.
2 With 10 percent plus 1 1/2 inch.
3 With no oversize.
4 Assuming labor at $3 per shift.
5 All material over 1 1/4 inches in size screened out at head box.
6 Stumps in gravel, and rock ledges above bedrock.
7 Only lower 18 inches of gravel washed.

Oroville.- Two men shoveled into boxes on the bank of Feather Rvier near Oroville, Calif., during the spring of 1932. The material handled in this operation consisted of loose top sand and small gravel on a river bar where gold had been concentrated recently. The gravel and sand were shoveled onto a grizzly with 1 1/2-inch openings at the head of the sluice. As the oversize piled up it had to be shoveled aside. The boxes were moved as required to keep the head of the sluice within easy shoveling distance of the gravel being mined. The gold was very fine, but apparently a good recovery was made. The water was pumped from the Feather River near by, using a 4-cylinder automobile engine.

Jones.- In June 1932 J. B. Jones and his three sons were mining a sloping bed of gravel on Mary Ann Creek, near Oroville, Wash. The gravel bench was about 100 feet wide. Cuts were taken at right angles to the creek. The boxes were set even with the surface, the end of the lower box was about 6 feet high at the edge of the flat bottom land.

A 5-inch centrifugal pump driven by an old 6-cylinder automobile engine lifted the water from a sump in the creek below the workings a vertical distance of 37 feet through 10-inch irrigation pipe to a point on the hillside, whence it was led by gravity to the head of the sluice boxes. About 200 gallons per minute was thus provided, without approaching the capacities of either the engine or pump. The water from the sluice box joined the very small flow of the creek and was again caught in the pump sump, having deposited most of its suspended load before reaching that point. The boxes were placed the full length of the cut before shoveling began. The first cut was 16 feet wide, with the sluice in the middle. (See fig. 6, A.) The boxes were supported with trestles as the gravel underneath was removed. Two men worked on either side of the box beginning at the lower end. The forward pair advanced a bench 3 1/2 to 4 feet deep, which included 2 feet of soil on top of the gravel but contained very few boulders that could not be washed. The lower bench consisted of 2 to 2 1/2 feet of gravel which was picked and shoveled into the sluice by the other two men. The boulders over the size of a man's fist were rolled or thrown back from the lower bench as the cut progressed upward. In succeeding cuts the boulders were thrown or rolled into the previous cut.

The riffles consisted of 8 feet of steel matting, such as is used for doormats, laid on coco matting and placed in about the middle of the sluice. In a test run of 50 cubic yards of gravel only 2 cents worth of gold was obtained in two sections of pole riffles placed below the steel matting, after which the poles were discarded. When the gravel contained clay, 1/2-inch cleats were placed ahead of the regular riffles at intervals of a few feet. These tended to break the clay as it went down the sluice.

At the start 12-inch boxes and correspondingly more water were used, but to reduce the consumption of gasoline for pumping the boxes were narrowed to 8 inches. With four men shoveling steadily the sluice box clogged occasionally above the riffles and had to be cleaned out. The tailings had to be shoveled away from the end of the sluice periodically.

Peshastin.- Four men were shoveling into boxes on Peshastin Creek, near Blewitt, Wash., in July 1932. They worked on the same side of the box and took the ground out in four benches. The last man cleaned bedrock as he came along. (See fig. 6, B.) In the first cut all boulders were lifted or tossed over the box. In the next cut the boulders would be rolled under the box into the space made by the previous cut. Two 7-foot sections of iron-clad Hungarian riffles were used for catching the gold. The remaining space in the boxes was covered with longitudinal pole riffles to protect the bottom of the box. Water was taken from the creek where an ample supply was available.

Minnick.- Two men were shoveling into boxes on Peshastin Creek above the operation just described. The gravel was full of large stumps, and rocky ledges projected into the gravel from the bedrock. If the time taken in cleaning the extremely uneven bedrock is considered not over 1 cubic yard could be handled per man-shift. Three 12-inch boxes were used. Riffles consisted of wire screen over carpet.

Blackhawk.- Two men were shoveling into boxes near Blackhawk, Colo., in July 1932. One man dug the gravel and shoveled it into a box 5 feet above the bottom of the pit at the head of the sluice box. He sorted out all material larger than a man's fist and threw it back of him in the pit. The second man pulled the gravel in a steady stream from the box over a 1/2-inch screen set at an angle of 45°. He also shoveled the oversize back into the completed part of the pit. The sluice was 10 inches wide; the riffles consisted of wire screen over carpet.

Greetan.- Two men at Bearmouth, Mont., were loading gravel into a 4 1/2-cubic-foot car which was dumped into a puddling box at the head of the sluice boxes. Only the lower 18 inches of the gravel was gold-bearing; the overburden was cast to one side.

The gold-bearing gravel, which contained about 10 percent of clay, was washed free of clay by hoeing it back and forth in the puddling box. The gate of the puddling box was raised when the box was full of water, and the mud and water and part of the gravel were allowed to run through the sluice boxes. The water supply was insufficient to wash the gravel through the sluice without the surges made by thus impounding it. The water was brought into the puddling box through a fire hose, and the small pressure available assisted in washing the gravel.

The washing arrangement and boxes were similar in design and operation to those used at some small-scale drift mines. The puddling box was made of matched lumber and was 15 feet long, 6 feet wide, and 2 feet high. The lower end was tapered to 8 inches, the width of the boxes.

DeWitt.- During the summer of 1932 H. DeWitt used a dry washer near Wilcox, Ariz. After the winter rains began he used a sluice with an end-shaking movement. The material washed was a red surface clay and gravel. After picking, it was allowed to lie exposed to the sun for a day or more. The drying caused the clay to slack so that it disintegrated in the sluice.

The sun-dried material was pushed to the head of the sluice in a wheelbarrow and fed into the box by hand. The sluice was 18 inches wide, 6 inches high, and 22 feet long and was set on rollers. The shaking movement was imparted by a rod from the head motion of an old Wilfley table. The stroke was about 5 inches and the speed 20 strokes to the minute. The rod connected to the end of the sluice had coil springs on both sides of the end board. The downward stroke was against stiff springs set on either side of the box. On release of the downstream pressure the side springs gave the box a quick flip backward.

The first 5 feet of the box was lined with a sheet-iron plate in which holes had been made with a pick. The projecting rugged edges of the holes tore up the gravel as it passed over them. Riffles consisted of 1- by 1-inch cross strips of wood 1 inch apart. Water was hauled 1 1/2 miles in barrels. About 1 cubic yard could be handled per man-shift. A cubic yard per hour could be washed in the box when dirt and water were available.

GROUND-SLUICING

General Features and Application

As previously defined, ground-sluicing is a placer-mining method in which gravel is excavated by running water not under hydraulic pressure. In some cases water is stored in reservoirs, and greatly increased flows are discharged during short periods. this variety of ground-sluicing is termed "booming" by placer miners and is so called in this paper. The action of the ground-sluice water may be augmented by hand work or by a jet of water under pressure. However, if hydraulic monitors or "giants" perform most of the work the method becomes "hydraulicking", described in a later paper.[78]

---

78 Gardner, E. D., and Johnson, C. H., Placer Mining in the Western United States: part II. - Hydraulicking, Treatment of Placer Concentrates and Marketing of Gold: Inf. Circ.6787, Bureau of Mines, 1934,

With a few exceptions ground-sluicing is used in small operations by miners with little capital. Otherwise it is used only where water under a substantial head is not available or where the gravel deposit is too small to justify building ditches or installing pipe lines for hydraulicking. The chief application of ground-sluicing is to deposits in or near stream beds. If applied to higher gravels pipe lines, flumes, or ditches are necessary, and the method loses one of its chief advantages, a low initial cost.

Ground-sluicing is not adapted to mining large deposits of tight or cemented gravel; these demand the tearing, disrupting force of giants. Gravel containing large boulders can be ground-sluiced, but more labor is necessary than if hydraulicking is used; the ground-sluice water seldom loosens and transports as large boulders as a giant or as large ones as will run through the sluice.

At many places no preliminary development work is necessary; at others long cuts must be run to reach bedrock under the deposit. If booming is to be done a dam must be built for a reservoir and perhaps an automatic gate. A sluice box is always used, although in many mines for little more than as a race to carry away the tailings. In bouldery ground derricks or drag lines may be needed to remove boulders.

A large number of small ground-sluicing operations were being carried on during the 1932 season by groups of 1 to 4 men working for themselves. Enough gold to pay regular wages was obtained at very few of the operations. Usually the miners received less than $1 per day for their efforts. A few larger operations have been carried on for a number of years; some of them have been profitable. Essential data concerning the larger mines and representative smaller ones operated in 1932 are shown in table 7. These mines are described in more detail later. They illustrate the applications of the method and general ground-sluice practices.

Types of Placer Deposits Worked

Except for the Osborne lease at Superior, Mont., gravels of moderate thickness along stream beds were being worked by ground sluicing at the mines listed in table 7. The depths of gravel ranged from 5 to 22 feet. Except at two placers very little of the gravel was tight, and none of it was cemented. At these, the Rundle and Morgan placers, Blackhawk, Colo., the gravel was tight and had to be loosened by picking; however, most of the material mined was overburden which was easily ground-sluiced.

Water Supply

The water supply was limited at most of the mines worked by ground-sluicing. At the Rundle and Morgan placers more water was available in the creek than could be used to advantage, because the gravel had to be loosened by picking. As noted in table 7, 60 to 500 miner's inches were used at the various mines listed. The natural flow of the streams ranged from 10 to 500 miner's inches.

Miner's inch.- Western placer miners usually measure and think of water in terms of miner's inches. In California and Montana, as established by law, 40 miner's inches equals 1 cubic foot per second; in Colorado the legal ratio is 38.4 to 1. Forty miner's inches to the cubic foot per second is generally accepted throughout the West; this value of the miner's inch is used in this paper. A miner's inch as used here equals 11.22 gallons per minute; 1 cubic foot equals 7.48 gallons.

TABLE 7.- Representative placer operations using ground-sluice method of mining, 1932

Name of mine	Mine			Gravel				Bedrock	
	Operator	Location	Address	Thickness, feet	Character	Percentage of boulders over 6 inches in diameter	Percentage of clay	Kind	Character
Morgan	Richard Leoncavallo.	Clear Creek	Blackhawk, Colo.	5	Tight, clay-bound	20	5	Clay	Soft.
Ravano	Tony Ravano	Harris Creek	Laurin, Mont.	6	Medium	25	0	Limestone	do.
Bar No. 1	...	Bar at mouth of Kamloops Creek.	Granite, Colo.	6	do.	10	0	False clay.	do.
Osborne	W. H. Osborne	California Gulch Cedar Creek.	Superior, Mont.	15	Fairly easy to pick.	15	0	Not on bedrock.	...
Rundle	W. B. Rundle	Clear Creek	Blackhawk, Colo.	8	Medium	15	2	Porphyry	Rough.
Bar No. 2	Jim Wiley	Bar at mouth of Kamloops Creek.	Granite, Colo.	6	do.	10	0	False clay.	Soft.
Kamloops	Kamloops Placer Gold Mining Co.	do.	do.	18	do.	10	0	Not on bedrock.	...
Willow Creek	Lawry, Kennedy, et al.	Willow Creek	Therma, N. Mex.	20	do.	10	1	do.	...
Camp Bird	Joe Witherspoon	California Gulch Cedar Creek.	Laurin, Mont.	10	do.	15	0	Limestone	Soft.
Bennet	S. B. Bennet	Quartz Creek	Rivulet, Mont.	15	do.	10	2	Porphyry	Hard, medium rough.
Harvey	C. W. Robertson	Sauerkraut Creek	Lincoln, Mont.	22	do.	35	5	...	Soft.
Magnus	Magnus & Ole Lindquist, Inc.	Swauk Creek	Liberty, Wash.	20	do.	25	1	Not on bedrock.	...

TABLE 7.- Representative placer operations using ground-sluice method of mining, 1932 - Continued

Name of mine	Water									
			Booms				Auxiliary cutting stream			
	Average miner's inches[1]	Capacity of reservoir, acre-feet	Water used, miner's inches	Average length minutes	Period between hours	Numbers per day	Head, feet	Length of pipe, feet	Diameter of pipe, inches	Diameter of nozzle, inches
Morgan	70	None	None	None	None	None	None	None	None	None.
Ravano	30	1	150	240	20	1	do.	do.	do.	do.
Bar No. 1	60	None	None	None	None	None	do.	do.	do.	do.
Osborne	80	do.	do.	do.	do.	do.	do.	do.	do.	do.
Rundle	60	do.	do.	do.	do.	do.	27	600	10	2.
Bar No. 2	60	do.	do.	do.	do.	do.	6	70	20 to 5	2.
Kamloops	100	do.	do.	do.	do.	do.	15	350	8-6-5	2.
Willow Creek	70	1 1/4	300	27	1 1/2	4	None	None	None	None.
Camp Bird	70	1/2	650	30	4	2	60	200	6	1.
Bennet	10	1/10	1,600	1 1/2	4	6	None	None	None	None.
Harvey	[11]300	5/8	7,500	2 1/2	1	24	22	20	18	
Magnus	500	1 3/4	4,000	15	1 3/4	14	None	None	None	None.

[1] 1 sec.-ft. = 40 miner's inches.
[11] Average 700, first 52 days.

1776

TABLE 7.- Representative placer operations using ground-sluice methodof mining, 1932 - Contiued

Name of mine	Handling material			Sluice Soxes					
	Percentage removed by hand or derrick	Maximum size of boulders put through boxes, inches	Method of handling boulders	Width, inches	Height, inches	Length, feet	Number	Grade: Inches per foot	Grade: Percentage
Morgan	40	6	Hand	18	12	10	2	1 1/5	10.
Ravano	40	3	do.			12	3	1/2	4.2.
Bar No. 1	15	4	do.	12	8	12	3	1 1/2	12.5.
Osborne	15	6	Steam derrick	[4]13	13	12	7	1/3	2.8.
Rundle	30	3	Fork and wheelbarrow	24	12	16	1	1	8.3.
Bar No. 2	15	4	Hand	18	12	12	2	1/4	2.1.
Kamloops	20	2	Power drag line with 3-in. grizzly buckets	30	20	12	42	1/4	2.1.
Willow Creek	10	6	Hand	18	10	12	50	1/2	4.2.
Camp Bird	20	3	do.	22	16	12	3	1	8.3.
Bennet	10	6	do.	36	18	12	3	3/4	6.2.
Harvey	30	9	Hand derrick	38	36	12	16	3/4	6.2.
Magnus	0	15		48	36	12	84	3/5	5.

[4]24 in., when more water was available

TABLE 7.- Representative placer operations using ground-sluice method of mining, 1932 - Continued

Name of mine	Riffles					Labor			Gravel washers, cubic yard			Costs Per cubic yard[2]		
	Type	Width, inches	Height, inches	Spacing, inches	Total length, feet	Number of men on each shift	Total number of men per 24 hours	Total number of 8-hr. shifts during season	During season	Per day	Per man-shift	Labor	Supplies	Total operating
Morgan	Wire screen over carpet.				20	1	1	60	165	2 3/4	2 3/4	$1.27	$0.02	$1.29.
Ravano	Pole	3	3	1	36	1	1	[3]84	775	9	9	.39	0	.39.
Bar No. 1	Wooden cross, over burlap.	1	1	1	36	3	3	100	300	9	3	1.17	.02	1.19.
Osborne	Pole	3	3	1	84	2 and 4	6	156	[5]1,400	53	9	.39	.03	.42.
Rundle	Cross, on wire screen over corduroy	3/4	3/4	5	16	1	1	60	174	3	3	1.17	.02	1.19.
Bar No. 2	Wooden cross, over burlap.	1	1	1	24	2	2	33	400	24	12	.28	.03	.31.
Kamloops	Wooden cross.	1 1/4	1	1	250	2	4	220	[6]4,000	73	18	.20	.04	[7].24.
Willow Creek	Wood block, round.	18	5	0	600	4	4	260	[8]1,030	16	4	.87	.04	.91.
Camp Bird	Pole	3	3	1	36	2	2	[9]134	2,400	36	18	.20	.02	.22.
Bennet	do.	3	3	1	36	1	1	[10]69	1,150	17	17	.21	.01	.22.
Harvey	do.	6	6	1	96	1	1	120	3,800	32	32	.11	.03	.14.
Magnus	20-lb. steel rails.	1 1/3	2 5/8	2	400	9	9	805	[12]6,000	67	7	.50	.04	.54.

[3]Includes 12 shifts cleaning up. [5]To July 8. [6]To July 17. [7]Not including rental of shovel, or supervision. [8]To July 20. [9]Includes 50 shifts cleaning up. [10]Includes 19 shifts cleaning up. [12]To June 23.

Sluicing

Operations at all ground-sluicing mines listed in table 7 consisted of running cuts preliminary to mining or mining as contrasted to developing.

Ground-sluicing has an advantage for running bedrock cuts in that a relatively small capital expenditure is required to test the gold content of the gravel and to ascertain the mining conditions. Cuts made in this manner also open up the deposit for mining. A relatively low yardage per shift is obtained in such work. A large part of the expense of running such cuts is the construction of the sluice, which will be used if the deposit is mined later. The work at the Rundle, Osborne, Kamloops, Willow Creek, and Magnus mines, at the time the authors visited them, consisted of running bedrock cuts.

At the other seven properties strictly mining operations were under way. Regular ground sluicing can be subdivided further according to the manner in which the water is used: (1) Water poured over the upper face of the cut and (2) water directed along the bottom of a straight or curved bank. Under (1) the cascading effect of the water is utilized to break down the gravel, or the fall of the water from a projecting flume or pipe is used to cut the bottom layer. Under (2) the current is used to undercut the bank and carry away the broken gravel at the face as it breaks down. The depth of the gravel has an important bearing on the practice followed; with thick deposits the cascading or falling of the water over the upper face is likely to give the best results; in thinner gravels the other method is usually advantageous. The width of the deposit is also a factor; in narrow channels the first method would have an advantage over the other. As shown by table 7 the thickness of gravel at the Morgan, Bar No. 2, and Harvey placers where the water was brought over the face was 5, 6, and 22 feet, respectively. At the other four placers, Ravano, Bar No. 1, Camp Bird, and Bennet the depth of gravel was 6, 6, 10, and 15 feet, respectively.

Of the 12 mines listed in table 7 sluicing with the natural flow of the stream was followed at 3. At 5 properties a minor part of the water was used under pressure, and booming was practiced at 5 not counting the Ravano. In one placer (Camp Bird) a hose was used for cutting the bank and the broken gravel removed by booming.

The water may be directed against the bank in a channel made by piled boulders or by boards. It can be diverted conveniently to any desired point or direction by means of what are called shears. These consist of 1- or 2-inch boards 12 feet long nailed to two tripods which slope back from the water flow at an angle of about 60°. The tripods are of pointed poles about 4 feet long and so constructed that boulders may be piled on the base to hold them in place. A row of "shears" may be used to direct the force of the water against a bank, or two rows may be used to form a flume.

Sluicing With Natural Flow of Stream

Of the multitude of "snipers" working on the creeks in the summer of 1932 many washed their gravel by ground-sluicing with the natural flow of the stream. The only equipment essential for this type of work in shallow gravels is a few 12-foot sluice boxes and picks and shovels. Table 7 shows that 2 3/4, 9, 3, and 9 cubic yards, respectively, were handled per man-shift at the first four mines in the table where the natural flow was utilized. At the Morgan placer water running down the face of a narrow cut was used to carry the fine material in the gravel to the box after it was loosened by picking. At the Ravano a cut was made first down one side of the area to be washed during the season and then extended across the channel. The water was directed against the bank by means of dams. Although a reservoir was used and the stream of water was augmented for a half shift each day the mine is placed under this class, as it is representative of a large number of operations where only

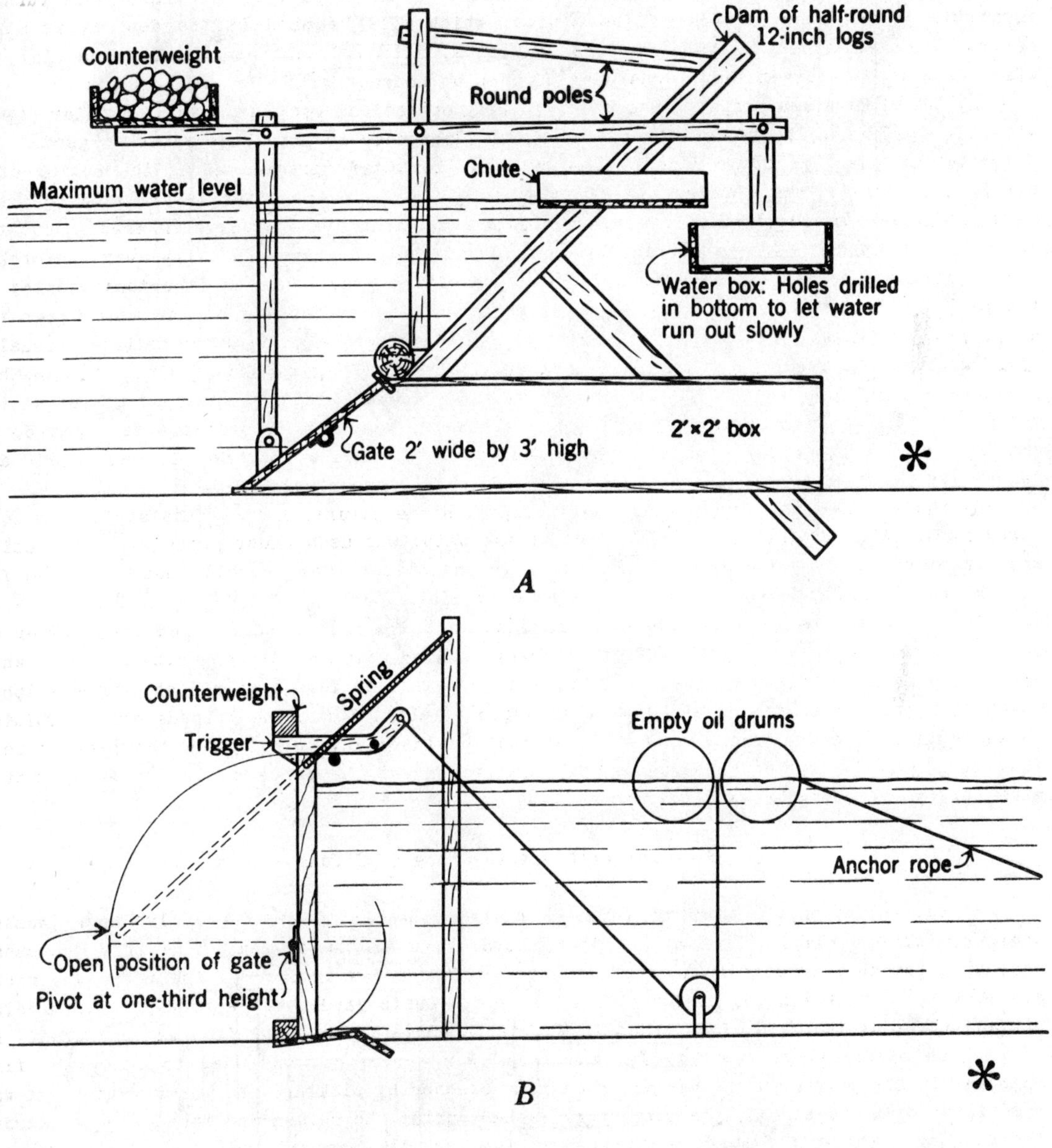

Figure 7.—Two types of automac gates for booming: *A*, Gate used at Bennet mine, Rivulet, Mont.; *B*, gate used at Harvey mine, Lincoln, Mont.

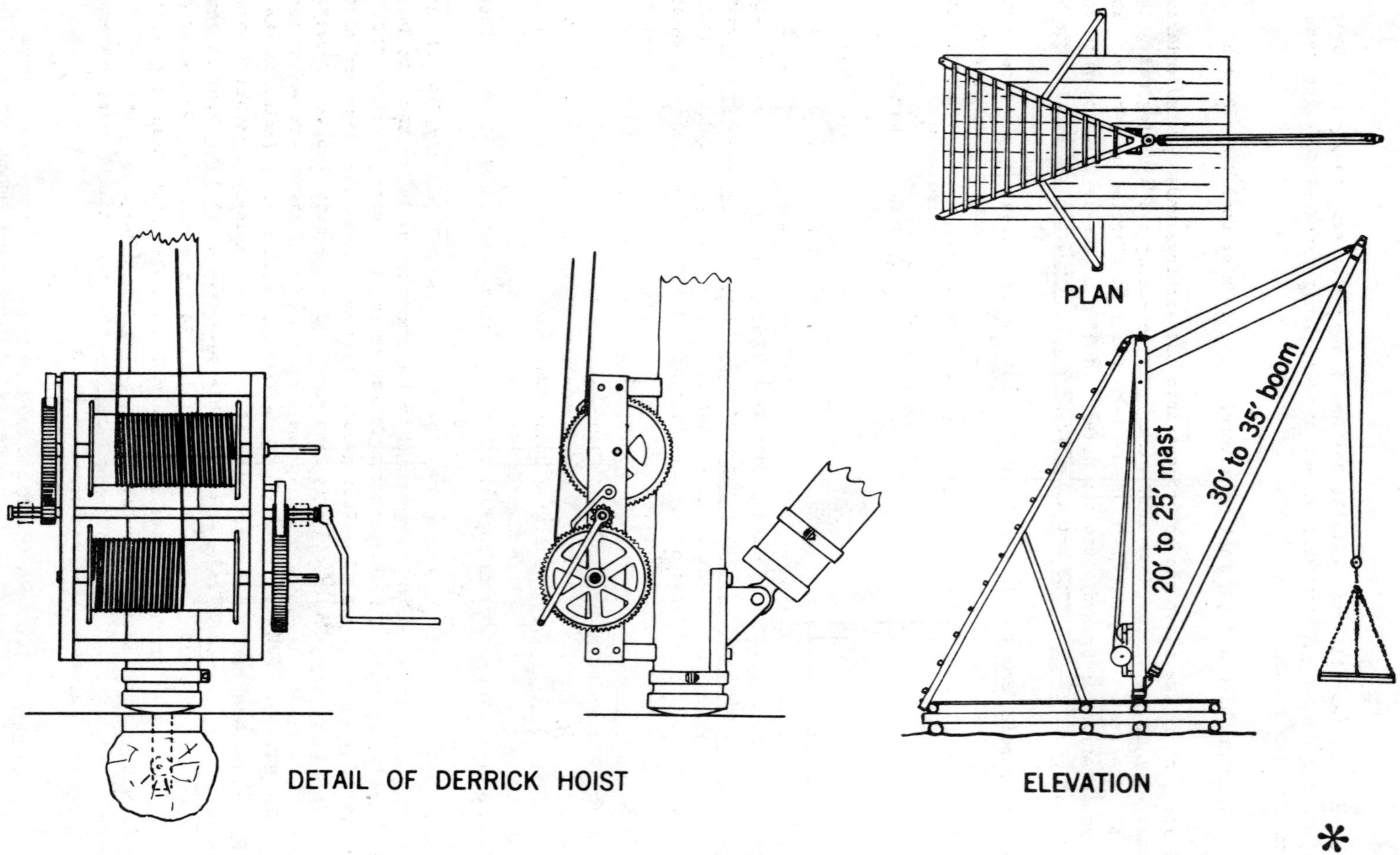

Figure 8.—Derrick used at Harvey placer, Lincoln, Mont.

the natural flow was utilized. At the Bar No. 1 the water was brought down over the face. At the Osborne deeper gravels were mined, and the water cascaded down the face of a cut. Boulders were removed by a power derrick.

Sluicing With a Minor Part of Natural Flow of Water Used Under Pressure

Where part of the water can be brought into the workings under pressure through a pipe the nozzle may be used to replace hand-picking in loosening and disintegrating the gravel. It may also be used to replace part of the hand labor in cleaning up bedrock. The ground-sluice water washes the gravel into the boxes; it also breaks down the bank to various degrees at different mines. The actual sluicing operations are quite similar whether or not water under pressure is used.

At the Rundle, Bar No. 2, and Kamloops mines, where pressure pipes were used the head of the water ranged from 6 to 60 feet, as shown in table 7; nozzles 1 and 2 inches in diameter were employed. The yardage moved per man-shift was 3, 12, and 18, respectively.

Booming

Booming utilizes the increased cutting and transporting power of water under flood conditions. As pointed out, water is stored in a reservoir and then released, flowing for relatively short periods. At the beginning of the season, when high water prevails, the booms may occur frequently, and each one may last relatively long. As the supply of water fails booms occur less often until finally there is not enough water to operate. Booming is the most important type of ground-sluicing. A much larger duty can be obtained per unit of water by booming than by other forms of ground-sluicing. The increased volume of water carries boulders into the sluice that otherwise would have to be moved by hand or by power and breaks down the banks against which a smaller stream would be ineffective. Larger sluice boxes must of course be used when booming than when utilizing only the natural flow of the stream. Booming is used in running development cuts as well as in strictly mining work. Of the three strictly mining operations listed under this head a side cut was used in two and an overcast at the end of the pit in the third.

The average number of booms per day at the five mines (excluding Ravano) ranged from 2 to 24. The duration of booms was 1 1/2 to 30 minutes.

The duration of a boom is governed by the size of reservoir and the flow of water. The capacity of the reservoir should be governed by the character of the ground. In heavy, rocky ground a short period with a correspondingly larger surge is more effective than a longer period with a smaller stream. In other ground, such as that at the Camp Bird mine, longer periods of washing, with correspondingly less water, give the best results. Reservoirs usually are constructed by building a dam across a narrow part of the stream bed or canyon and backing the water up behind it. Ordinarily, earth dams with a board facing on the upstream side are used. Reservoirs with capacities of 1/2 to 1 3/4 acre-feet (table 7) are used at the mines listed. Automatic gates have a double advantage in that no labor is required to operate them and the booming in the pit can continue 24 hours each day. Automatic gates are shown in figure 7. Although the water is not as effective when unattended as when the miners are on shift it accomplishes considerable work during off hours.

Table 7 shows that 4, 18, 17, 32, and 7 cubic yards, respectively, were washed per man-shift at the mines where booming was done.

Sluice Boxes and Riffles Used in Ground-Sluicing Mines

Table 7 also shows that where only the natural flow was utilized boxes ranged from 12 to 30 inches in width and where booming was used from 18 to 48 inches in width.

Wooden cross riffles were used at 4 mines where ground-sluicing was done with the natural flow of the stream. Pole riffles were used in 2, and wire screen over carpet was used in 1 other. Corduroy under wire screen was used under the riffles in 1 and burlap in 2 others.

More substantial riffles are required in the boxes where the gravel is boomed on account of the coarser material and greater volumes put through the sluices. At 3 mines where booming was done pole riffles were used, at 1 mine wooden blocks, and at 1 mine 20-pound steel rails.

The general subject of sluice boxes and riffles is discussed in more detail in a subsequent paper.[79]

Handling Boulders

Boulders generally are handled in ground-sluice mines in the same manner as in hydraulic pits. Derricks are the commonest mechanical device used for the purpose. (See fig. 8.) These may be operated by a hand winch or by steam, gasoline, or electric power. In wide deposits boulders may be removed on platform-skips operated on a cableway. At the Harvey placer, where the ground contains an unusually high percentage of boulders (35 percent over 6 inches in diameter), two hand-operated derricks are used in a pit 90 feet wide. At the Osborne lease a steam-operated derrick is employed. At the other ground-sluice mines listed boulders are moved by hand; large ones are blasted. At the Osborne lease large boulders are drilled with jackhammers, using steam from the boiler which supplies power for operating the derrick. Boulders too large to move by hand are blown from the pit by explosives at the Camp Bird mine.

Bringing the ground-sluice water over the face of the pit has a decided advantage in that boulders need to be handled only once. They are either moved completely out of the workings or piled in the pit on cleaned bedrock, as is done at the Harvey placer. Where a side cut is taken, as at the Camp Bird mine, boulders are moved back over the washed area as washing progresses across the channel. Small stones are tossed free of the pit on the side from which sluicing started. If the bedrock is cleaned only at the end of the season boulders must be moved again.

Cleaning Bedrock

As previously stated, at end-cut mines bedrock is cleaned as the work progresses upstream. Usually a section of the pit corresponding to the length of a sluice box is cleaned at a time. Many miners consider the amount of gold cleaned each box length as the unit of value of the ground. With a side cut the bedrock usually is cleaned at the end of the season. Although frequent clean-ups bring in a current revenue the practice of cleaning up at the end of the season permits full advantage to be taken of the water while it is available. Moreover, cleaning at the end of the season affords an opportunity for a more nearly complete recovery of the gold, as the work can be done in a more leisurely manner.

---

79 Gardner, E. D., and Johnson, C. H., Placer Mining in the Western United States: Part III. - Dredging and Other Forms of Mechanical Handling of Gravel, and Drift Mining: Inf. Circ. 6788, Bureau of Mines, 1934, 81 pp.

Bedrock had not been reached in 4 out of 5 places where cuts were run for this purpose. In putting in new boxes, however, layers of gravel on the bottoms of the pits were shoveled carefully into the sluice to save any gold that had been left behind during the washing operations.

Where the bedrock is soft a layer is removed and washed as in other forms of placer mining. The depth taken depends upon how deeply the gold has penetrated into it. The bedrock material may be washed into the main sluice box or shoveled into special clean-up boxes. In mines where a side cut is used a separate clean-up box usually is employed. Advantage is taken of a hose in cleaning bedrock when one is available.

Where bedrock is hard it must be cleaned by hand in all ground-sluice operations. Even if the surface of the bedrock is smooth it almost invariably contains soft seams and cracks which must be dug out with small hand tools.

The sluice-box concentrates usually are panned or washed in a rocker, as described under Separation of Gold and Platinum from Concentrates in a later paper.[80]

Descriptions of Ground-Sluicing Mines Operating in 1932

The mines described are the larger ones and a few typical small ones visited by the authors in 1932. Comparable data concerning them were shown in table 7.

Morgan.- Richard Leoncavallo was working a pit on the Morgan placer on Clear Creek below Blackhawk, Colo., in July 1932. The gravel was tight, coated with clay, and overlain by 2 or 3 feet of recent wash and mill tailings. Cuts about 6 feet wide radiated from the head of the sluice boxes, following rich streaks on a false clay bedrock. All the gravel had to be loosened by picking, which was done while the water was running over the face of the cut. Boulders more than 6 inches in diameter were thrown out by hand. Some of the gravel near the head of the sluice was shoveled in by hand.

About 70 miner's inches of water was used; some of the sediment in the creek water was settled out in a small reservoir above the mine. The sluice consisted of two 10-foot boxes 18 inches wide and 12 inches deep having a grade of 12 inches per box. The first 4 feet was floored with 1-inch screen placed tight on the bottom of the box. Below this 1/2-inch screen was laid over carpet a and canvas. A canvas apron was placed in the pit just ahead of and a few inches below the level of the sluice; most of the gold was caught on this canvas. The clay bedrock was cleaned by shoveling the top layer into an 8-inch box built and set up for the purpose.

An average of 2 3/4 yards was washed per day with one man working. At $3.50 per day the labor cost would have been $1.27 per cubic yard. The total operating cost, allowing $0.02 for supplies, would have been $1.29 per yard.

A number of men were conducting similar operations farther down Clear Creek; none were making wages.

Ravano.- About 775 cubic yards of gravel was sluiced by Tony Ravano in 3 months, including 12 days for cleaning up bedrock, on Harris Creek near Laurin, Sheridan County, Mont. The average run-off in the creek during the season was about 30 miner's inches. This water was stored overnight in a small reservoir, the gate of which was partly opened in the morning, allowing a stream of about 150 miner's inches to run until noon. Only the natural flow ran during the afternoon. As the flush water reached the pit it was deflected by boards and rock dry walls against the bank on one side of the pit and thence through a sluice. (See fig. 9, A.) The action of the water was assisted by picking into the bank. During the

80 Gardner, E. D., and Johnson, C. H., Placer Mining in the Western United States: Part III. - Dredging and Other Other Forms of Mechanical Handling of Gravel, and Drift Mining: Inf. Circ. 6733, Bureau of Mines, 1934, 81 pp.

afternoon boulders over 6 inches in diameter were rolled back onto the bedrock previously exposed. Stones 3 to 6 inches in diameter that were not carried out by the water were thrown clear of the pit by hand. All fine material was washed through three 12-foot boxes with longitudinal pole riffles, but nearly all of the gold stayed on the bedrock. Other miners working on the same creek by this method did not use any boxes except for cleaning up; they considered that a good extraction of the gold was made in this manner.

In clearing up bedrock the sluice was extended into the pit one box at a time. The boulders were removed from the bedrock, which was then loosened by picking, and the material was shoveled into the box. As each box was placed the boulders from the next section above were moved back on the part cleaned up where they formed a bed about 2 feet deep. All of the material that was piled on bedrock had been handled twice. A total of 84 shifts was worked, and an average of 9 cubic yards was moved each day. At $3.50 per shift the labor cost would have been 40 cents per cubic yard. There were no expenses other than labor as old timber was salvaged for the sluice boxes.

Bar No. 1.- In July 1932, Bar No. 1 placer on Kamloops Creek near Granite, Colo., was being mined by three men working on one shift per day. About 60 miner's inches of water was used, which came from a ditch at the head of the area being worked. Three 12-foot boxes 12 inches wide and 8 inches high, with a grade of 18 inches to the box, were used for a sluice.

The ground washed was 6 feet deep, comprising a flat bar on a false clay bedrock. There were very few boulders. Conditions were favorable for ground-sluicing. The men, however, were inexperienced and did not work to the best advantage. The water was spread out over too much ground and irregular cuts were taken, leaving small tracts that could be moved only by hand. The boulders were handled 2 or 3 times before final disposal. Gold was lost, as not over one half of the bedrock area was cleaned up.

About 9 cubic yards was handled daily. With wages at $3.50 per day the labor cost would have been $1.17 per cubic yard. Allowing 2 cents for supplies the total cost would have been $1.19.

The adjoining ground was being worked by an experienced man who moved nearly four times as much ground per man-shift with the same amount of water and extracted a larger proportion of the gold in the gravel.

Osborne.- In July 1932, W. H. Osborne with three other men was running a ground-sluice cut in a gravel bench on California Gulch of Cedar Creek near Superior, Mont. This cut was to reach bedrock and was then 120 feet long and averaged 20 feet wide and 15 feet deep. The water cascaded down the face of the cut. Two men loosened the gravel with picks and rolled boulders out of the way. One man worked on the flume and one built boxes and assisted elsewhere when needed. Two men worked on night shift. Small boulders were piled beside the sluice. The ground contained some large boulders which were removed from the cut every other day by a steam derrick working half a shift. Those up to 3 tons in weight could be hoisted with the derrick. Larger ones were drilled with a jackhammer, using steam from the derrick boiler, and blasted.

The water supply was insufficient for good work. On July 8 about 80 miner's inches was available. The boxes had been 24 inches wide, but as the water decreased the width was narrowed to 13 inches. The grade of 4 inches per 12-foot box was too flat for the coarse material being mined; the boxes clogged if not attended continually. Table 7 shows that 9 cubic yards was being washed per man-shift. At $3.50 per shift the labor cost would have been 39 cents per cubic yard. Supplies cost about 3 cents, making a total of 42 cents.

Rundle.- W. B. Rundle was ground-sluicing at Blackhawk, Colo., in July 1932. Part of the water used was from a pressure pipe. The material being washed consisted of 2 or 3 feet of tight creek-bed gravel overlain with 5 or 6 feet of recent wash and mill tailings. A cut was started at the rim of the creek bed and extended upstream; the face had just reached bedrock. The layout of the workings is shown in figure 9, B. The ground was partly loosened, and the boulders were washed clean by water from a 2-inch nozzle under a head of 27 feet. Some picking was necessary to loosen the virgin gravel. The overburden and the loosened material were washed into the boxes by the ground-sluice water which poured over the end of the pit.

All material over 3 inches in size was forked out into a wheelbarrow and taken to a rock dump. The ground sloped steeply upward on the hill side of the pit, and the oversize had been forked out of the pit on the creek side until the rock pile reached such a height that it was easier to use a wheelbarrow.

The riffles consisted of wooden cleats placed 5 inches apart and were of 3/4-inch square material with the downstream edge beveled backward. The bottom of the box was lined with corduroy cloth with the corrugations at right angles to the box; wire screen was placed over the corduroy. The water contained flotation tailings, including a large quantity of pyrite from a concentrator up the creek. The sulphide, however, did not clog the riffles in the sluice box.

An average of 3 cubic yards was washed per day. At $3.50 per shift the labor cost would have been $1.17 per cubic yard; the total cost, with 2 cents per cubic yard for supplies, would have been $1.19.

Bar No. 2 - In July 1932, Jim Wiley, with one helper, was mining a flat grass-covered gravel bar near the Arkansas River at the mouth of Kamloops Creek, Colo. The gravel occurred on a false clay bedrock which sloped about 1 inch to the foot. A 6-inch sand streak occurred just above the bedrock. The ground was worked in cuts 8 feet wide and 150 feet long. There was enough dump room at the lower end of the cuts. The bank was undercut in the sandy layer by a jet of water from a fire hose, using a 2-inch nozzle and a 6-foot head; the overlying gravel broke down and was washed into the boxes by the ground-sluice water. About 60 miner's inches was used. Lumps were broken by picking, and all boulders were thrown back beside the sluice boxes. A tight wing dam built of sod directed the water into the head of the boxes. A 5- by 20-foot steel sheet was placed in the cut 18 inches below the level and ahead of the sluice; the edges of the sheet were turned slightly upward, and the space underneath was packed at the edges with sod. The sheet was moved up to the face of the cut as soon as room was made; the sluice was extended to the lower end of the sheet at the same time. Bedrock was cleaned up by shoveling the top of the false clay bedrock onto the sheet where clay that did not break up was puddled with a hoe or shovel. The material on the sheet was then shoveled into the boxes. Most of the gold was caught on the sheet.

The boxes were 18 inches wide and had a grade of 1/4 inch to the foot. The riffles consisted of 1- by 1-inch transverse wooden cleats placed over burlap. Quicksilver was used in the boxes when cleaning up.

About 16 cubic yards per man per day was washed when ground sluicing. Including clean-up time, the average was about 12 cubic yards per day. At $3.50 per day the labor cost would have been 28 cents. The total operating cost, including 3 cents for supplies, would have been 31 cents per cubic yard.

Kamloops.- The Kamloops Placer Gold Mining Co. was running a cut on a bench to reach some reputed rich ground near Granite, Colo., in July 1932. The cut started at the edge of old hydraulic workings. Bedrock had not been reached. Part of the gravel was tight and contained some boulders 4 or 5 tons in weight. On July 17, 1932 a cut 15 feet wide and 525 feet long had been run. The average depth was about 14 feet and the maximum 18 feet. The sluice was 30 inches wide, 20 inches high, and 504 feet long. Dredge-type Hungarian riffles were used. Boxes were set at the flat grade of 3 inches to the 12-foot box to reach bedrock as soon as possible. To prevent clogging, as much of the washed oversize as could be loaded conveniently was removed from the pit ahead of the boxes with a power dragline. This machine had a 35-foot boom and a 40-foot line. Boulders up to 3 tons in weight could be lifted out of the cut with the dragline by using chains. Boulders over 3 tons in weight were blockholed and blasted. The dragline bucket or dipper was made with a grizzly bottom with 2-inch spacing between the bars. About 20 percent of the material was removed by the dragline.

Water under a 15-foot head was directed against the bank by a 2-inch nozzle mounted on a stand with a "gooseneck." The ground-sluice water flowed down the face. About 100 miner's inches was used.

The crew consisted of two men. One operated the dragline and nozzle, the other attended the sluice and picked down the bank. The dragline operator on day shift acted as superintendent. Two shifts were worked, and an average of 73 cubic yards was handled per day. Allowing a wage of $3.50 per day for all four men the labor cost would be 20 cents per cubic yard. With supplies at 4 cents per cubic yard the total operating cost would be 24 cents, disregarding supervision, rental, and repairs to the dragline.

Willow Creek.- Four men - Laury, Kennedy, Lund, and Neal - were running a cut to reach bedrock on Willow Creek near Therma, N. Mex. On July 20, 1932 the cut was 130 feet long and averaged 24 feet wide and 10 feet deep. A dam had been built above the mine across the creek on bedrock, which raised the underground flow of water above the surface. When ready to boom the gate in the dam was opened by hand. The water poured out of a 10-inch pipe undera a 3-foot head and was conducted in a 12-inch pipe to the head of the cut. The pipe extended over the face of the cut, and the stream of water struck the toe of the slope. The pit was boomed four times per day; the length of each booming period was 27 minutes. The rest of the time, while the reservoir was refilling, was spent in throwing out boulders and installing boxes. While booming, two men worked in the face with shovels, assisting the action of the water. The third man watched the boxes to see that they did not become clogged, and the fourth man stayed at the end of the sluice box and pulled away the tailings when they tended to pile up.

The pit had not reached bedrock; when this occurred the face would be widened to include all of the gravel channel. The sluice was 18 inches wide, 10 inches high, 600 feet long, and set on a grade of 6 inches to the 12-foot box. The riffles consisted of round blocks 5 inches long and 18 inches in diameter that just fit in the boxes. A 1- by 4-inch strip nailed on the inside of the box held the blocks in place. In addition to acting as gold catchers the blocks protected the bottom of the boxes.

Lumber cost $22 per M at a sawmill near by. An average of 4 cubic yards per man-shift was being washed; not enough water was available for the economical utilization of the labor. At $3.50 per shift the labor cost would have been 87 cents per cubic yard. With supplies at 4 cents, the total operating cost would have been 91 cents. A large part of the work consisted of building 470 feet of sluiceway down the canyon, below where the cut was started, to provide dump room. Deducting the cost of this, the labor cost would have been about 40 instead of 87 cents.

Camp Bird.- Joe Witherspoon, with one man, was ground-sluicing on California Gulch near Laurin, Sheridan County, Mont., during the 1932 season. The gravel occurred along the bottom of the gulch on the present stream course and averaged about 7 feet deep and 20 feet wide. The pit, an extension of old workings, was 450 feet long on July 5, 1932. (See fig. 9, C.) Overlying the gravel was 2 1/2 feet of loam. The top soil was piped off with a 1-inch nozzle on a firehose connected to a 6-inch pipe with a 60-foot head. The hose also was used for cutting the bank. The pit was boomed on an average of twice a day, the flush water running about 30 minutes each time. The remaining time was used in removing boulders and cutting the bank. Enough water was available in the early part of the season to boom 4 or 5 times per day. The ground-sluice water was deflected and held against the bank by boards and dry walls. A part of the natural flow of the stream ran through the pit while the boulders were being cut and thrown out. Boulders over 6 inches in diameter that could be lifted by hand were rolled or lifted back on the washed bedrock. Stones 3 to 6 inches in diameter were thrown by hand clear of the pit. Boulders too large to lift were blown out of the pit with 40-percent-strength gelatin dynamite. Several sticks of explosive were placed under the boulders so that on detonation the boulders were lifted clear of the pit. The boulders were blasted when partly submerged in water, which acted as stemming for the explosive. Three 12-foot boxes, 22 inches wide and 16 inches high, set on a grade of 1 inch to the foot were used; little gold, however, got into the sluice.

It was estimated that about 50 shifts would be required to clean up the bedrock that was exposed on July 5. In doing this the boulders would have to be moved a second time. The bedrock was soft and would be picked and shoveled into clean-up boxes.

Between 50 and 60 cubic yards was being washed per day. Allowing clean-up time, the average for the period was about 36 cubic yards per day or 18 per man-shift. At $3.50 per day the labor cost would have been 20 cents, and with 2 cents per cubic yard for supplies the total operating cost would have been 22 cents.

<u>Bennet</u>.- I. B. Bennet had been ground-sluicing by booming on a side gulch of Quartz Creek near Rivulet, Mont., for a number of years. The bedrock sloped about 1 1/2 inches to the foot. The gravel was 12 to 25 feet deep and averaged about 15 feet. As the reservoir filled it discharged through a 24- by 24-inch opening, the gate of which operated automatically. Figure 7, A, shows the details of the automatic gate. At the beginning of the season the reservoir filled every 2 hours and for a period of 9 days, when the snow was melting most rapidly, every 20 minutes. At the end of the season the reservoir filled only once a day. The flush water was directed against the bank, which was kept in the form of a semicircle. As the bank was washed away, the coarser gravel, about 3 feet in depth, was left on bedrock. At the end of the season this 3 feet of material was loosened by picking, and the boulders were rolled back onto bedrock; the water carried away the rest of the material. An area 20 feet wide and 100 feet long was washed during the 1932 season of 50 days. About 1 foot of boulders was left to be moved again as the bedrock was cleaned. It was estimated that about 3 weeks would be required for this work. The bedrock was hard and medium rough. It would be cleaned by hand; crevices must be dug out and scraped. An average of 23 yards was mined per 8 hours during the washing period. Although only one 8-hour shift per day was spent on the ground, the booming went on for 24 hours.

At $3.50 per day the labor cost would be 21 cents per cubic yard. The total operating cost was about 22 cents.

<u>Harvey</u>.- C. W. Robertson had been working the Harvey placer near Lincoln, Mont., for the past 15 years. The gravel worked was in the bottom of a gulch along the stream bed. The depth ranged from 18 to 24 feet and averaged 22 feet; the width was 90 feet (fig. 9, D). The gravel contained an unusually large proportion of boulders, some of which weighed as much as 8 tons. The pit advanced the full width of the channel. The washing was done by booming. When work began in the spring the reservoir would fill in 9 minutes, and the full stream would run for 15 minutes. In July a boom occurred once an hour and ran only 2 1/2 minutes. The gate opened automatically (see fig. 7, B). The water was taken from the reservoir in a flume 5 feet wide, 3 feet deep, and 400 feet long. At the head of the pit the flume divided into five branches, the ends of which extended over the face of the pit. During booming the water dropped into the pit through two adjoining branches. The bottom 6 feet of the gravel was fairly tight and partly bound with clay. The overcast allowed the water to drop on this stratum; as the lower gravel was cut away the bank above sloughed down. The sluice was 38 inches wide, 36 inches deep, and 192 feet long and had a grade of 3/4 inch to the foot. Individual boxes were 12 feet long. Pole riffles 12 feet long were used in the boxes. The sluice was cleaned up at the end of the season. It was brought upward in the pit to one

side of the center (see fig. 9, D). Twelve-foot cuts were taken from the face of the pit on alternate sides of the sluice. A new box was put in after every pair of cuts.

Boulders were handled by means of two well-built hand-operated derricks and piled back of the ground being worked; they filled the pit to a depth of 10 feet. A dry wall was built up of boulders on either side of the sluice. One derrick stood on a platform across the head of the sluice boxes on the rock walls. The other derrick was on a platform on the main rock pile. The derricks were built of 10- to 12-inch round timber. The boom of the one over the sluice box was 30 feet long and that of the other 35 feet long. The masts were 20 feet long. The winch cables were 5/8 inch in diameter. The derricks were set so that a load from one could be taken up and dumped by the other. A sling platform was used for handling any boulders that a man could roll onto it. A chain was used for larger ones. No blasting was necessary.

Each 12-foot section of bedrock was cleaned up as the cut was completed. The bedrock, which was fairly soft, was loosened by picking and then hosed into the sluice by a 3 1/2-inch firehose connected to an 18-inch pipe from one of the boxes on the bank. The 22-foot head did not give enough pressure to cut the bedrock. After a section was cleaned a team was brought into the pit, and as many large boulders as could be handled from near the face were dragged onto the cleaned-up bedrock. A dry wall was then built to deflect the water to the head of the sluice for mining the next cut. Old rags were used in the wall near the head of the box, and sod was used elsewhere to make the wall watertight for a height of 12 or 18 inches. The wall was raised and the space back of it filled with boulders as the next cut was taken out above.

To July 7, 1932 when the water had materially decreased, 2,520 cubic yards had been washed in 50 working days. A total of 3,800 cubic yards was washed by the end of the season (in the middle of September). Robertson worked the mine alone. Up to July 7, 50 cubic yards had been washed per day. Although only one shift per day was worked the booming went on 24 hours. The labor cost would have been 7 cents; allowing 3 cents for supplies the total cost was 10 cents per cubic yard. For the whole season an average of 32 cubic yards was washed per shift. This would amount to 11 cents per cubic yard for labor, or a total of 14 cents. Some work, such as repairing the dam and flume, was done during the winter; this would increase the total cost per cubic yard 2 or 3 cents.

Magnus and Ole Lindquist, Inc.- This company was running a long cut by ground-sluicing to reach bedrock near Liberty, Wash. After bedrock was reached it was planned to put in hydraulic equipment. On June 23 the cut was about 400 feet long and averaged 20 by 20 feet in cross-section. The sluice boxes were 48 inches wide and 36 inches high. The grade was 5 percent. Riffles were 20-pound steel rails set lengthwise 2 inches above the bottom of the box and 2 inches apart. All boulders up to 15 inches in diameter were put through the boxes. Any over this size were first broken by blasting, then run down the sluice.

Water for booming was stored in a reservoir; the gate was 6 by 6 feet and was opened automatically. The boom lasted 15 minutes, and an average of six booms occurred during the shift; the booming continued 24 hours each day. The water poured over the face of the cut; it broke down the face and scoured out the cut without much assistance from the miners. Boulders occasionally were started rolling and assisted through the boxes by hand.

Of the 9 men employed 3 worked at the sawmill cutting lumber for the sluice boxes; 3 were putting new boxes in the cut as room was made and at the lower end as the tailings filled up the limited dump room; 2 worked at the end of the sluice, leveling off the tailings; and 1 man watched the boxes and supervised the work. The three men who worked on boxes also handled boulders and watched the water during the booming period. The low yardage (7 cubic yards) per man-shift was due mainly to the unusually large number of men employed in cutting lumber and putting in boxes. The total labor cost at $3.50 per shift would have been

50 cents. With 4 cents for supplies the total cost would have been 54 cents per cubic yard. Exclusive of the extra men required to level off the tailing and to extend the boxes at the lower end of the sluice on account of limited dump room, the labor cost would have been about half that indicated.

Summary of Ground-Sluicing Operations and Costs in 1932

Although in the past ground-sluicing has been conducted on a fairly large scale, all operations using this method of mining in 1932 were relatively small. The largest number of men per day (4, 6, and 9) were employed where ground-sluicing was used for developing gravel deposits. Only 1, 2, or 3 men were employed at mines where strictly mining operations were being conducted. An average of 32 cubic yards per day was mined by one man at the Harvey mine, although the gravel contained the unusually high proportion of 35 percent of boulders over 6 inches in diameter. At the other ground-sluicing mines the daily yardage handled ranged from 2 3/4 to 73 and the cubic yards per man-shift from 2 3/4 to 18. The total yardage handled during the entire 1932 season at the mines listed was about 43,000. Although a large number of small mines not listed herein were being worked in 1932 the total yardage moved was not great. If all of them were considered the 43,000 cubic yards listed probably would not be trebled.

The cost of ground-sluicing like that of other methods of placer mining varies directly with the cubic yards handled per man-shift. As shown in table 7 the labor costs per cubic yard for each of the four mines ground sluicing with natural flow of stream were $1.27, $0.39, $1.17, and $0.39, respectively. Supplies were estimated to cost up to 4 cents per cubic yard. The labor costs at the mines using an auxiliary hose with the natural flow were $1.17, $0.28, and $0.20, respectively. Supplies were estimated to cost 2, 3, and 4 cents per cubic yard. Labor costs at five mines where booming was practiced were 87, 20, 21, 11, and 50 cents, respectively. Supplies at the same mines were estimated at 4, 2, 1, 3, and 4 cents per cubic yard.

No supervision or overhead charges were incurred at any of the mines except the Kamloops and Magnus and Ole Lindquist where cuts were being run preparatory to large-scale work. In these two mines working superintendents were employed, and their time at the regular wage rate is included in the operating cost; the charges for supervision would be the salary paid in excess of $3.50 per day.

As stated, the cost per cubic yard of gravel moved in development work was high because of the large proportion of the work involved in placing boxes. Since little equipment was used at most of the mines the amortization charge would be low. The following table shows the estimated total operating costs at the mines where strictly mining operations were being carried on by ground-sluicing.

Mine	Daily yardage	Variation of method	Operating cost per cubic yard
Morgan......	2 3/4	With natural flow..........	$1.29
Ravano......	9	do.	.39
Bar No. 1	9	do.	1.19
Bar No. 2	24	With auxiliary nozzles	.31
Camp Bird	36	Booming..............................	.22
Bennet......	17	do.	.22
Harvey......	32	do.	.14

At the Morgan placer the gravel was loosened with difficulty by picking, hence the high cost per cubic yard. The low efficiency of labor at the Bar No. 1 mine was mainly the result of inexperience. The costs at the other mines listed represent average conditions.

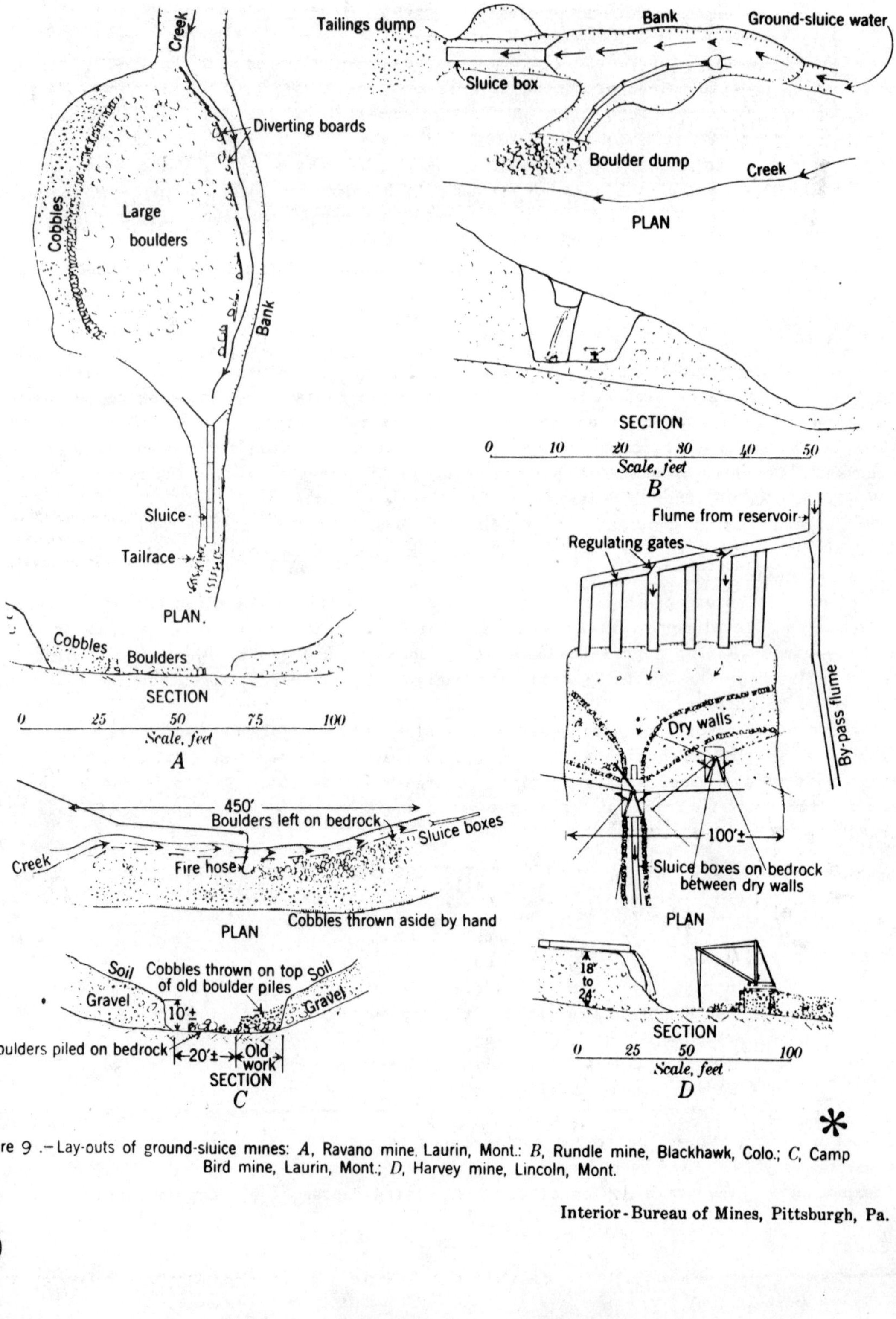

*

Figure 9 .–Lay-outs of ground-sluice mines: *A*, Ravano mine, Laurin, Mont.: *B*, Rundle mine, Blackhawk, Colo.; *C*, Camp Bird mine, Laurin, Mont.; *D*, Harvey mine, Lincoln, Mont.

Interior-Bureau of Mines, Pittsburgh, Pa.

GOLD/SILVER WEIGHT CONVERSION TABLE

	GRAIN	GRAM	DWT TROY	OZ TROY	LB TROY	OZ AV	LB AV	CARAT
GRAIN	1.0	0.06479	0.041667	0.0020833	0.00017	0.0022887	0.0001428	0.3240
GRAM	15.4324	1.0	0.06479	0.03215	0.0027	0.03527	0.002205	5.0
DWT TROY	24.0	1.5517	1.0	0.05	0.00416	0.0548571	0.0034285	7.7755
OUNCE TROY	480.0	31.10346	20.0	1.0	0.0833	1.09714	0.06857	155.51
POUND TROY	5780.0	373.241	240.0	12.0	1.0	13.1657	0.82286	1866.12
OUNCE AV	437.0	28.3495	18.2297	0.9114883	0.07595	1.0	0.0625	141.75
POUND AV	7000.0	453.592	291.667	14.5833	1.21107	16.0	1.0	2267.96
CARAT	3.168	0.20	0.03215	0.0064304	0.000536	0.007055	0.000441	1.0

Abbreviations: AV = Avoirdupois, DWT = Pennyweight, LB = Pound, OZ = Ounce

DIRECTIONS: To find conversion values, find basic weight in left-hand column and follow this line until it intersects with the desired weight. For example, if you desire to determine how many 'ounces Troy' there are in a 'pound Troy', locate POUND TROY in the left-hand column and follow it across until you intersect with OUNCE TROY and you will find '12.0, which is the correct number of ounces Troy in one pound Troy. Then, if you want to know how many ounces there are in 2½ pounds Troy, multiply 12.0 by 2½ and arrive at a correct figure of 30 ounces Troy. Another valuable way to use this table is for comparative purposes: for comparing avoirdupois against Troy, for example. If you want to compare the relative actual weight of a pound Troy against an avoirdupois pound, follow the POUND AV line across to the POUND TROY column and you can determine that a pound avoirdupois is equal to 1.21107 pounds Troy. Using the same method, on the other hand, you will find that an ounce avoirdupois is equal to 0.9114883 ounce Troy. Anyone who buys or sells gold or silver will find hundreds of valuable applications for this table which are not immediately apparent to the casual user.

You are invited to reproduce this conversion table for distribution to club members or in other publications, provided that credit is given to the Placer Miner's Manul, Volume 1. Free copies are available by sending a stamped self-addressed envelope with a request for the table to: Exanimo Press, Box 18, Segundo, Colorado 81070, or to the National Prospector's Gazette, Miner's Hall, Ames, Nebraska 68621.